Praise for *The "God" Part of the Brain*

"I very much enjoyed the account of your spiritual journey and believe it would make excellent reading for every college student–the resultant residence-hall debates would be the best part of their education. It often occurs to me that if, against all odds, there is a judgmental God and heaven, it will come to pass that when the pearly gates open, those who had the valor to think for themselves will be escorted to the head of the line, garlanded, and given their own personal audience."

–Edward O. Wilson,
two-time Pulitzer Prize–winner

"An impressive compilation of data and ideas…both accurate and thoughtful."

–E. Fuller Torrey, MD ("The most famous psychiatrist in America"–the *Washington Post*)

"All six billion plus inhabitants of Earth should be in possession of this book. Alper's tome should be placed in the sacred writings section of libraries, bookstores, and dwellings throughout the world. Matthew Alper is the new Galileo…Immensely important…Defines in a clear and concise manner what each of us already knew but were afraid to admit and exclaim. The cat's out of the bag."

–John Scoggins, PhD

"A lively manifesto...For the discipline's specific application to the matter at hand, I've seen nothing that matches the fury of *The "God" Part of the Brain*, which perhaps explains why it's earned something of a cult following."

–Salon.com

"This is an essential book for those in search of a scientific understanding of man's spiritual nature. Matthew Alper navigates the reader through a labyrinth of intriguing questions and then offers undoubtedly clear answers that lead to a better understanding of our objective reality."

–Elena Rusyn, MD, PhD, Gray Laboratory,
Harvard Medical School

"Your book was sensational. Your writing was clear and concise; your summation was bold and masterful."

–William Wright, author of *Born That Way: Genes, Behavior, Personality*

"Vibrant...vivacious...an entertaining and provocative introduction to speculations concerning the neural basis of spirituality."

–*Free Inquiry* magazine

"Thank you for making sense out of the hunches and gut feelings I've had for years. I feel more peaceful and positive now. I hope that the candle you've lit in the vast darkness will burn as bright as the sun."

–John Emerson, PhD

"The best in its field…brilliant."

–Noe Zamel, MD, FRCPC

"Mr. Alper has written an extremely readable and comprehensive analysis of the physiological basis of religiosity…comparable to Freud's *Future of an Illusion* in its contribution to the continuing maturation of the human mind. I am using *The "God" Part of the Brain* to teach a Sociology of Religion course with remarkable results."

–William Dusenberry, PhD

"I greatly appreciated *The "God" Part of the Brain* as it so nicely summarized and integrated much of the work being done in this field."

–Andrew Newberg, MD, PhD,
author of *Why God Won't Go Away*

"Matthew Alper is high maintenance. Not only is his intellect superior to most PhD candidates that I know, but his intensity in displaying that intellect and arguing his world view is more compelling than many of my grad school courses. So, here I am, fiercely advocating this unconventional, first-time author who, with one slim book, has thrown hundreds of years of human religious beliefs out the window and replaced them with a concise scientific view of spirituality that is impossible to argue with. The brain is the secret. In our brains lie nature's survival mechanisms in which God is nothing but a protective lens through which humanity is 'programmed' to view the world. Matthew Alper has the chutzpah to remove that lens, to crush it under his heel, and then, as we cringe in the unfiltered light, he dares us to look up and stare into the pure scientific truth he has discovered. *The "God" Part of the Brain* is a challenge at first, but once you open your mind to the potentials of its theories, there is nothing to do but follow its arguments to their logical conclusions. And although he rips away our old stiff crutches, this audacious philosopher is kind enough to spoon-feed us a new and positive way to approaching our existences."

–Rebecca Morris, Editor-in-Chief,
Cardozo Law Journal

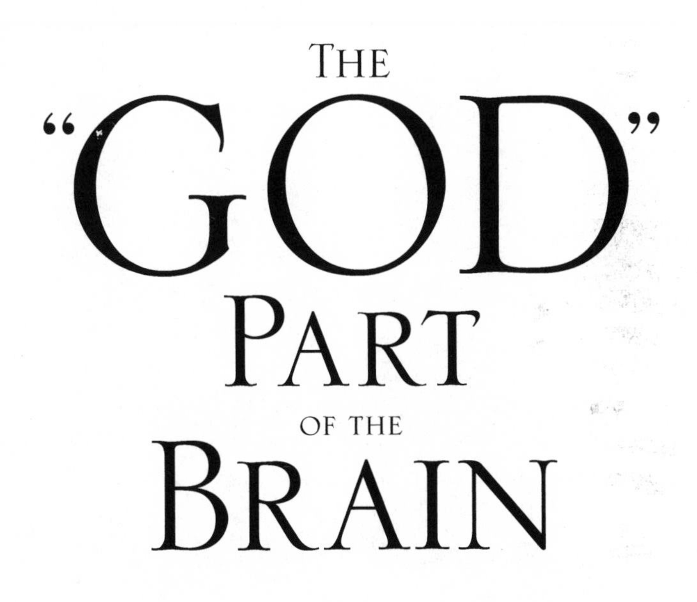
The
"God"
Part
of the
Brain

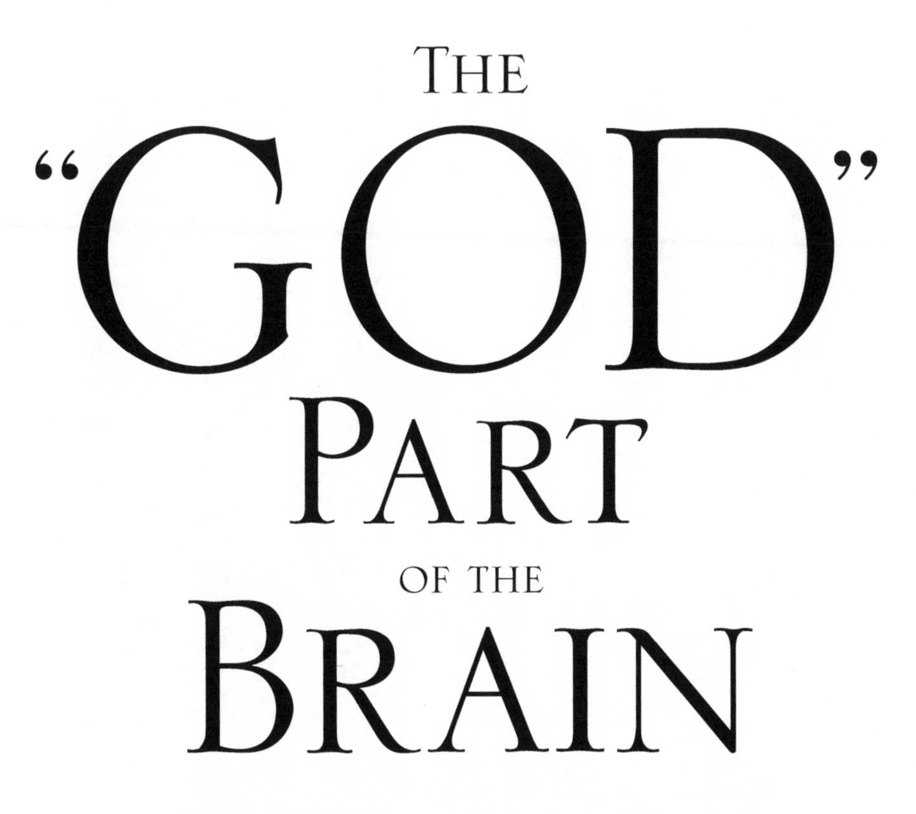

THE "GOD" PART OF THE BRAIN

A Scientific Interpretation of Human Spirituality and God

MATTHEW ALPER

SOURCEBOOKS, INC.®
NAPERVILLE, ILLINOIS

Published by Sourcebooks, Inc.
P.O. Box 4410, Naperville, Illinois 60567-4410
(630) 961-3900
Fax: (630) 961-2168
www.sourcebooks.com
Originally published in 1996 by Rogue Press

Alper, Matthew.
The "God" part of the brain.
p. cm.
Includes bibliographical references and index.
1. Psychology, Religious. 2. Brain–Religious aspects. I. Title.
BL53.A47 2006
200.1'9–dc22
2006011690

Printed and bound in the United States of America.
POD 18 17 16 15 14 13 12 11 10

For more information go to: www.godpart.com
To write the author: godpart@aol.com

Acknowledgments

I would like to thank my parents, Joan and Jud, and sister, Elizabeth, for their enduring support; Dr. E. Fuller Torrey and Dr. Arthur Rifkin for fixing me; Tonya Bickerton-Watson for her invaluable time; John Stern; Art Bell; Lisa Lion; Edward O. Wilson; Helena Schwarz; Susan Rabiner; Sherry Frazier and Lisa Vasher at McNaughton & Gunn; Arnold Sadwin; William Wright; Joe Fried; Rebecca Morris; Albert Fernandez; Brandon Quest; Lori Wood; Daniella Monticello; Dominique Raccah; Hillel Black; Tara VanTimmeren; Matt Diamond; Megan Dempster; Genene Murphy; and all those innumerable others who have helped me along the way.

“GREAT IS THE TRUTH
AND MIGHTY
ABOVE ALL THINGS”
THE APOCRYPHA
I Esdras iv, 41

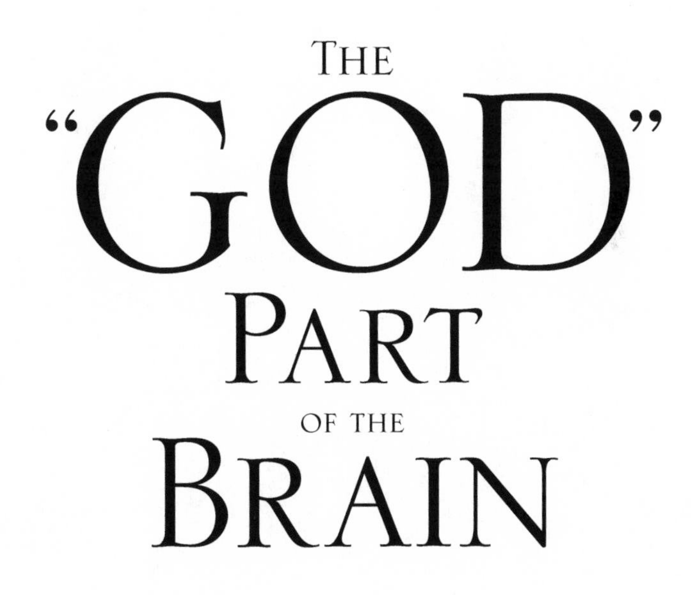
The
"God"
Part
of the
Brain

Contents

PROLOGUE

"Man finds himself in the world, or has been thrown into it, and as he stands facing the world he is confronted by it as by a problem which demands to be solved."

–NICHOLAS BERDYAEV

"I want to know God's thoughts...the rest are just details."

–ALBERT EINSTEIN

Knowledge is power, and it is precisely our species' capacity to reason–to deduce knowledge–that has secured us the title of "the most powerful creature on Earth." Human beings reason because we are compelled to do so. Our survival depends on it, for with every new piece of information we acquire, be it as individuals or a species, we become that much better equipped to master our world and therefore to survive.

In addition to this practical need to amass information, our species also seeks knowledge in the hope that it might provide us with a sense of meaning and purpose. In this regard, our species is unique from all certainty, it was that I was one day going to die. The question now was: Would death mark the decisive end of my existence or the advent of a new beginning?

Here I was at a time in my life when I was being asked to make such critical decisions as what career path I would take. Only how was I to concentrate on such trivialities with the problem of my own mortality left unanswered? How could I justify an interest in tomorrow while ignoring the greater question of where I would stand against all eternity?

Moreover, why, when God was supposed to be both all-good and all-powerful, was there so much pain and suffering in the world? Why would an all-powerful God allow for so much misery and injustice to prevail in his kingdom? Why would He make us so fragile, so mortal? In time, I found it difficult to believe in a God that was both benevolent and omnipotent. Instead, it seemed that God, if He existed at all, was either all-good but not very powerful, or else–even more disquieting–He was all-powerful but not very good.

Without answers to such ponderous questions, my future stood before me like a metaphysical brick wall. The universe began to take on the proportions of an unfathomable void, which, if not sated with knowledge of God's existence, I was beginning to feel would eventually consume me. I needed answers. I needed to know. Was this a world of magic and miracles, or wasn't it? I wanted to see if I could find some tangible, verifiable data that would either prove or disprove God's existence once and for all.

And so, like an Arthurian knight in search of his Holy Grail, I said goodbye to the conventional world and, instead, rode off alone into the vast dark forest of existence in search of an answer to that ultimate problem: Does God exist? I spent many years lost in those seemingly impenetrable woods, often despondent and despaired, thinking I would one day die there without ever having resolved a single thing.

But at last, I have returned...furthermore, with what I believe might be the answer.

Book I

Theory's Evolution

"To question all things; never to turn away from any difficulty; to accept no doctrine either from ourselves or from other people without a rigid scrutiny by negative criticism; letting no fallacy or incoherence, or confusion of thought step by unperceived; above all, to insist upon having the meaning of a word clearly and precisely understood before using it, and the meaning of a proposition before assenting to it; these are the lessons we learn from ancient dialecticians."

–John Stuart Mill

"The unexamined life is not worth living."

–Socrates

"According to the doctrine of chance, you ought to put yourself to the trouble of searching for the truth; for if you die without worshipping the True Cause, you are lost."

–Pascal

CHAPTER 1

THROWING ROCKS AT GOD

"The Caterpillar and Alice looked at each other in silence for some time; at last the Caterpillar took the hookah out of its mouth, and addressed her in a languid, sleepy voice.

'Who are you?' said the Caterpillar.

Alice replied rather shyly, 'I–I hardly know, sir, just at present–at least I knew who I was when I got up this morning, but I think I must have been changed several times since then.'"

–LEWIS CARROLL

By the time I was twenty-one, my quest for knowledge of God had taken several unexpected turns. In this time, I had searched the world's myriad religions only to find myself frustrated by a gamut of flaws and inconsistencies in all their logic. I had investigated the various paranormal phenomena only to encounter a trail of false claims and chicanery. I had experimented with the mind-altering effects of psychedelic drugs as well as transcendental meditation, only to undergo a series of distorted sense-experiences, none of which had brought me any closer to acquiring verifiable knowledge of any spiritual reality or God. As a matter of fact, if anything, they had only served to draw me farther away. This was due to the fact that while exploring the effects of LSD, I had a bad trip that led to a severe clinical depression compounded by a dissociative, depersonalization, and anxiety disorder. For a year and a half, I suffered this unfortunate state until, finally, with the aid of pharmacological drugs, I was restored to my previous, relatively healthy self.

Though it may have come at a very high price, I nevertheless managed to garner some extremely valuable information from this otherwise wretched experience, information regarding the nature of my allegedly immortal human soul.

According to the various belief systems (religions) I had thus far encountered, the human soul was supposed to be spiritual in nature, a fixed and permanent agent, unalterable and everlasting. Again and again, I was told that when I died, though my physical body would perish, "I"–the sum of my conscious experience, the essence of my thoughts and feelings, what was perceived as constituting my soul or spirit–would persist for all eternity. The fact, however, that my conscious self had been so drastically altered convinced me that there was no fixed or eternal essence in me.

Twice in a year and a half, I had undergone two complete transformations of my so-called eternal self. First, my conscious self was transformed into something other than it previously had been by psychedelic drugs. Then, a year and a half later, my original self was restored, this time by a drug known as a monoamine oxidase inhibitor (MAOI). But I thought consciousness was supposed to be conceived in spirit–fixed, eternal, immune to the influences of physical nature. If this were true, how was it that the core of my conscious experience had been altered, twice now, by ingesting physical substances? How was it that a combination of molecules–raw matter–could affect something as allegedly ethereal as consciousness, that which was supposed to represent my immutable, transcendental soul? To believe that matter could affect one's spirit, that it could impact upon the soul, would be the equivalence, it seemed, to believing that one could throw rocks at God. If spirits or souls truly existed, it would seem they should be impervious to material influence.

The fact that my conscious self–my allegedly immortal soul–was susceptible to the effects of chemical (physical) substances convinced me that human consciousness must be a physical entity governed by strictly physical processes. If this was true, then in order to gain a deeper understanding of the nature of consciousness–what I previously believed might constitute a soul–I would need to conduct an investigation into the nature of the physical sciences.

Up until this point, I always had the greatest respect for the physical/natural sciences. I was always impressed by their ability to rationally explain most any phenomena as well as to lead to the creation of tools and technologies that worked to make our lives easier. Whereas in the past, however, in which I had admired the sciences, I now revered them. Science had saved my life. I was indebted to it. God didn't save me. I didn't save me. Science, the tool of reason, had saved me. I was my own living proof that science worked. And so, the same faith that many placed in a god or religion, I now placed in science. Simply, it was a paradigm which brought verifiable results. Not that I didn't have faith in science before this. Every time, for instance, I flipped a light switch, one could say I had faith the lights would go on. The difference was that, whereas in the past I had taken my faith for granted, I was now a staunch believer.

As I saw it, science had resolved the riddle of the human soul. Science had proven it could come up with chemical formulas that could manipulate the contents of one's cognitions, emotions, and perceptions in almost whatever way it saw fit. It could electrically or chemically stimulate parts of one's brain in such a way that it could make one passive or aggressive, tranquil or manic, happy or sad. In essence, science could alter and manipulate one's cognitive and emotional states as if pulling the strings on a marionette.

As a result, I was now convinced that the mind, which I previously believed to constitute my transcendental soul, instead represented the workings of my physical organ, the brain. There was no soul. There was no ghost in the machine. My thoughts–human consciousness–were not the manifestation of some ethereal force or will but rather the consequence of synaptic transmissions, electrical and chemical signals being registered throughout my brain, generating a host of sensations, perceptions, emotions, and cognitions in me–pure neuromechanics. Consequently, as far as I was now concerned, the riddle of the human soul had been solved. From hereon, I would interpret the origin of all perception, sensation, emotion, and cognition from a strictly neurophysiological–that is, scientific–perspective.

As secure as I now was that there was no such thing as a transcendental soul, I still found myself plagued by that more essential problem of God's existence. As God supposedly constituted the embodiment of all things spiritual, not until I possessed some rational explanation through which I could resolve the problem of His existence could I be absolutely certain there was no such thing as a transcendental/spiritual reality. And as long as it was possible that God might exist, it was therefore also possible that I possessed a transcendental soul. Consequently, before I could commit to anything, I needed to resolve the greater and all-encompassing problem of God.

As the physical sciences had helped me to rationally interpret the underlying nature of consciousness, I now wondered if it would be possible to apply this same tool of reason to resolve that ever-persistent problem of God. Could the physical sciences crack that nut as well? Up until now, it hadn't come close. From biologists to astro- and quantum physicists, no one had ever advanced anything resembling a scientific interpretation of God. But why was this? Did God truly exist only beyond our grasp, beyond the range of human comprehension? Or was there a physical solution, only no one had discovered it yet?

As a now firm believer in the methods of science, I felt there must exist a rational explanation for everything. As a scientific idealist, I found myself inclined to believe that nothing was beyond our reach. If it could be dreamt of, it could be reasoned through.

My course was now defined. I would be a scientist. I would accumulate all the scientific knowledge I possibly could and then, once this was accomplished, once I had familiarized myself with all the various disciplines, only then could I justifiably recommence with my quest for knowledge of God.

But wait! What if it should turn out that science was just another form of psychological indoctrination, a new religion for a new world? Granted, the fruits of science had helped me out of a dark depression, but what if it was just my faith in science that healed me, the result of some sort of placebo effect, no more or less valid than when one's maladies are cured by a religious faith healer? What if science was no more founded in truth than any of the other self-glorified creeds I had thus far encountered? Perhaps scientists were just the high priests of a new faith, one that, instead of referring to gods, referred to particles that were just as incomprehensible and elusive. Perhaps science was just another disingenuous paradigm, a new mythology for the modern age. Then again, perhaps it was not. Perhaps science was a genuine tool by which human beings could gain a clearer and more distinct insight into the underlying nature of reality. So which was I to believe? How could I prove that scientific facts were any more reliable than religious ones? It was time to define my terms, time to investigate the investigator. Before I would blindly place my trust in the scientific process, before I submitted myself to a lifelong quest for a scientific interpretation of God, I would first have to investigate the nature of my newfound faith. "What," I had to ask, "is science? How does it work?"

Chapter 2

What Is Science?

"Science is the attempt to make the chaotic diversity of our sense-experience correspond to a logically uniform system of thought."

–Einstein

"There is no such thing as absolute certainty, but there is assurance sufficient for human life."

–John Stuart Mill

In order to justify my quest for a scientific explanation of God, I first had to conduct an investigation into the nature of science itself. This is what I found:

What is science? Since this is a rather large question, I will do my best to explain it in the most conceptual terms I am able. Before I begin, however, let me state that no matter how much faith one places in science, he must realize that at no time can it ever represent anything more than just another belief system, just another way by which humans can choose to interpret reality. I say this not out of any lack of conviction but only because not even science can guarantee anything with absolute certainty. Nothing can! Who, for instance, could say with total assuredness that his experiences are anything other than an illusion or a dream? As written over two thousand years ago, "Once upon a time, I, Chang-Tzu, dreamed I was a butterfly, fluttering hither and thither when suddenly I was awakened. Now I do not know whether I was a man dreaming I was a butterfly, or whether I am a butterfly now dreaming that I am a man." Nothing is certain! No wonder one of the wisest men to walk the Earth, Socrates, lived by the principle that all he knew was that he knew nothing at all.

Nevertheless, with that necessary qualifier aside, let's presume for the moment that this experience we call life isn't a dream. Let's suppose for the moment that we do exist as, more or less, what we imagine and that our experiences are, for the most part, "real." Even so, it is still impossible for us to ever possess absolute knowledge of anything. Let me elaborate.

The only means we, as human beings, have to interpret reality is through information acquired through our physical sense organs. Through our eyes, we absorb photons of light; we see the world. Through our ears, we absorb vibrations; we hear it. Through the nerve endings that cover the surfaces of our skin, we experience differences in pressure and temperature; we feel the world. Through our noses and tongues, we absorb chemicals; we smell and taste it. Before we acquire knowledge of our world, all information must first pass through these physical sense organs. Consequently, our sense organs play a critical role in determining the manner in which we perceive reality. As each species possesses its own unique set of sense organs, each must therefore experience and, consequently, interpret reality from its own unique and relative perspective.

Common houseflies, for instance, have a different mechanism from ours by which they absorb light–they possess a different set of organs that we would call eyes. As flies sense the world differently from us, they must consequently interpret it differently. Just as a fly sees the world from its own unique fly perspective, we see the world from our unique human perspective. Whereas flies possess fly knowledge, humans possess human knowledge. And just as a fly can only possess fly knowledge and no other, a human can only possess human knowledge and no other. We must therefore accept that our interpretation of reality is no "better" or more "real" than a fly's. It's simply different.

Moreover, it's not just the manner in which our physical sense organs absorb information that determines our perspectives of reality but, just as significantly, the manner in which our brains then process that information. For instance, what does it mean when we say that we "see" an apple? First, photons of light which are reflected off an apple are picked up by our retinas, which convert that information into electrical signals that are then processed by our brain. Consequently, all that we perceive as "real" is nothing more than electrical signals as they are interpreted by our organ, the brain. When we eat an apple, we "feel" its texture; we "smell" its aroma; we "taste" its flavor. Not until we integrate all of these various sense-impressions is our experience transformed into a coherent perception of the apple as a whole. Without such an internal processor through which to coordinate this medley of sense-impressions we constantly receive, it would be impossible for us to make sense of our experiences.

In the least sophisticated organisms, such internal processors constitute a single neural pathway. As life evolved, so did this single pathway into an integrated neural network that converges at a central location called a ganglion. A more complex version of the ganglion, we call a brain. Ours, the human brain, represents the most sophisticated processor of all. Because each organism possesses its own unique processing mechanism, its own central nervous system or brain, each organism must therefore interpret reality from its own unique and relative perspective.

Furthermore, it's not just the different species that perceive and interpret reality from their own unique perspectives but also each individual within each species. Among our own species, each individual possesses his own unique combination of sense organs–his own unique combination of ears, eyes, nose, mouth, and skin. In other words, no two humans have the exact same set of sense receptors. For example, because the physical mechanics of my eyes are slightly different from my neighbor's, I will experience the color red differently than he does. In an even more extreme example, someone with damaged cone receptors, who is totally colorblind, will consequently experience what I perceive as bright red as toneless or gray. Because each individual perceives the world from his own unique perspective, each of us must consequently maintain our own unique interpretation of reality.

Just as each individual's sense organs vary, so does each individual's processor or brain. Just as no two people possess the same exact eyes, no two people possess the exact same brain. Therefore, not only does each individual acquire sensual data differently, but each of us then processes and therefore interprets that same data in his own unique way.

In addition to these factors, we must also take into consideration the fact that each individual lives a unique set of life experiences. As this, too, will impact upon one's cognitive development, it also affects the manner in which one will interpret reality.

There are therefore three variables that determine the manner in which each species (as well as each individual within each species) interprets reality. These include the physical nature of an organism's sense organs, the physical nature of its processor (brain), and the content of its life experiences.

With these three variables in mind, let's imagine that two amoebae, two houseflies, two chimpanzees, and two humans are all perceiving the same sunrise. As each of these individual entities absorbs and then processes the sun's radiated light energy in its own unique fashion, who could possibly say which of their experiences is the most authentic or "real"? What organism could dare claim that it sees the "real" sunrise? Which organism could say that its experience of the rising sun's red color is any more genuine? Red is a man-made construct that bears no relation to the actual physical universe, nor to the reality of other species. Though we may interpret the sunrise as being red, the sunrise "in itself" is not. This is just the manner in which the mean of our species experiences a particular wavelength (six hundred nanometers) of light as it falls upon our retinas. In essence, we must recognize that we can only conceive of reality inasmuch as our biologies enable us to do so.

As each of us perceives the world from our own unique and therefore relative perspective, all knowledge must consequently be relative as well. In the words of Immanuel Kant, it is impossible to know "things in themselves" but rather only "things as we perceive them." Consequently, it's impossible for us to ever know anything with absolute certainty. Instead, we can only know things with relative certainty. But if this is true, one might justifiably ask: Why seek to know anything at all?

The answer to this is simple. Regardless of how relative our perspectives might be, we nevertheless possess the capacity to perceive a close or common enough approximation of things as to provide us with practical information regarding our world. This is why, for instance, if we were to take a roomful of people all looking at the same rock and we were to ask them what they saw, though each individual might experience the rock from his own unique perspective, each will generally agree that the object at hand is indeed a rock. If, among this same roomful of people, some claimed to see a shoe, some a banana, others a dog, we'd be in for some trouble. Fortunately for our species, however, this is not the case. Our sense organs are consistent enough that if we were to place an object such as a rock in front of a roomful of people, the majority will generally agree that it is a rock they are perceiving. Though we may never know a "thing in itself"–though we may never possess absolute knowledge of anything, our perceptual organs and internal processing mechanisms offer us a consistent enough account of the world to provide us with practical and reliable data. As a matter of fact, our perceptual organs have yielded so much practical and reliable data that we have been able to develop entire scientific disciplines from them. These disciplines have helped us to cultivate such practical and reliable technologies as the electric light, microwave ovens, nuclear energy, artificial organs, spaceships, antibiotics, electron microscopes, and computers, to name a small few.

So what is science's secret? How does it allow us to take our perceptions of things and transform them into an electric light or microwave oven? What application of knowledge is this that it has furnished us with such a vast wealth of life-enriching technologies? Simply speaking, how does science work?

Science relies on a very strict process known as the scientific method, a process whose principles were originally outlined by two philosophical contemporaries, namely Sir Francis Bacon (1561–1626) in his book *Novum Organum* and Rene Descartes (1596–1650) in his book *Discourse on the Method of Properly Conducting One's Reason and of Seeking the Truth in the Sciences.* Descartes suggested that in order to procure what he referred to as "clear and distinct" knowledge of things, one had to apply a strict set of guidelines to the manner in which he conducts his observations. Descartes referred to these guidelines as the scientific method. And what is this scientific method? Without providing a detailed explanation of Descartes' own principles, I will attempt to offer a more conceptual interpretation.

The scientific process operates in two phases: the empirical and the statistical. In the first phase, a scientist seeks patterns in the universe based on empirical observation–data received through the physical senses. For example, based on information acquired through his sense organ, his eyes, an early human happens to notice the sun rising from the east. The next morning, he notices the same thing occur. After several more observations, this nascent scientist begins to recognize a pattern. Based on his initial observations, he may surmise that perhaps the sun, as a rule, rises from the east. Since he has yet to confirm this "theory," his assertions are, for the time being, purely hypothetical. After all, a few simple observations are hardly any basis for placing unconditional faith in something.

It is now, in the second phase of the scientific method, that our scientist must perform a series of tests that will either verify or refute his original hypothesis. He might, for instance, decide to observe the sunrise for several more years, allowing each morning's observation to represent one more piece of evidence to confirm his theory. This is where the statistical phase enters the picture.

After our scientist feels confident that he has obtained sufficient statistical evidence to support his theory, he will disclose his findings to those around him, more specifically to the rest of the world's scientific community. It is now the duty of the scientific community to review his hypothesis by performing their own series of tests. This is necessary as the conclusions of one sole observer should never be accepted as adequate proof of anything. What if, for instance, our original scientist was making up the results just to get attention or perhaps he was simply too ignorant to know the difference between east and west.

It is at this point that other scientists will perform their own tests meant to either confirm or invalidate the original scientist's findings. Perhaps some of these scientists will duplicate the original scientist's experiments to see if they get the same results. Others, meanwhile, may devise whole new means of testing the theory. One, for instance, may wish to see whether or not he will obtain the same data from some other part of the globe. Perhaps in Africa or Asia the sun rises from the west.

As this process continues, one by one, our ever-skeptical scientific community will conduct as many tests as they can come up with before assenting to a theory. Only after a sufficient amount of supportive statistical data is obtained might the scientific community be willing to give credence to a theory–in this case, that the sun does indeed rise from the east.

Keep in mind, statistics still do not reflect certainties. Though the sun may have consistently risen in the east for as long as humankind has recorded this phenomenon, the supposition that the sun rises from the east is still just a theory. Just because the sun has risen in the east every day up until the present doesn't necessarily mean that it will do the same tomorrow. How, for instance, can we know with absolute certainty that the sun won't explode this evening for reasons beyond our knowledge? We don't. What we do know is that the sun has been rising in the east for so long and with such consistency that it most probably will do the same thing tomorrow–not certainly, just most probably. Even Einstein recognized that though no one single experiment can ever prove a theory correct, all it takes is one to prove a theory incorrect. (For example, should the sun rise from the west, just once, there goes the entire theory.) Scientists do not therefore claim to be able to "see" into the future but only to predict within a certain degree of accuracy, based on probabilities, what may or may not occur.

But if science is based on mere probabilities (as opposed to certainties), why should we place so much faith in it? Why practice science with such conviction? The reason is that although the whole of science may be based on probabilities, it still represents the most accurate and reliable source of information any method, system, or paradigm has offered us thus far. Though our local meteorologist may sometimes provide us with an inaccurate forecast, how often do we choose to turn to our local priest, shaman, or psychic for tomorrow's weather? Though scientific method may be based on mere probabilities and therefore imperfect, it has proven itself, time and time again, to represent the most reliable and accurate source of information we have.

Once the scientist has probable cause to give credence to a theory, once he has faith that the pattern he has recognized occurs with a sufficient degree of consistency, he will then use this newfound information to elicit even more. One deduced "fact" can be used to deduce the next. Once our scientist accepts that the sun rises from the east, he is now armed with yet one more fact with which to decipher his universe, one more piece of the puzzle with which to try to grasp the greater picture. In his search for answers, the scientist will utilize his findings to uncover even more elusive patterns. In this way, science is constantly building upon itself.

One of the fundamental principles of science is that every action has an effect. This, in turn, suggests that every effect has its cause. Once a theory has been verified, a scientist might want to know why such a thing occurred. Once he accepts, for instance, that the sun rises in the east, he may want to dig deeper into the mystery of this phenomenon by asking: Why does it rise this way? Is it because a sun god is pulling it up from the east by a magical string or maybe because the Earth revolves around the fixed Sun from that direction? Presuming that the sun rises from the east, the scientist may now search for yet an even deeper understanding of this phenomenon.

With the assistance of various tools that can be used to enhance our empirical powers of observation (e.g., a telescope with which to augment our vision), a scientist can dig perpetually deeper into the mysteries of the physical universe, acquiring information one piece at a time until he has acquired as much knowledge as is humanly possible.

Now there are those who refute the scientific method, those who deny its capacity to reliably interpret our world, those who consider it a sham, an artifice, a means of deceit. They refer to science as the Devil's plaything, a conspiracy developed to contradict their own religious beliefs. Take, for instance, those who support the Judeo-Christian interpretation of the Earth's origins, otherwise known as creationism. Such "creationists" reject man's evolution from the primates. They reject the idea that the Earth (as well as life) is a few billion years old. Regardless of how much their beliefs (e.g., that the world was created in six days approximately six thousand years ago) may contradict libraries full of carefully documented scientific data (data acquired through the exact same methodology that gave us the electric light and automobile), they insist that their viewpoint is correct. How is it that such people can refute such well established data and yet, in the same breath, turn on their electric fans when they are overheated or take antibiotics when they are ill? How can people spurn the sciences one day and then gladly partake of their fruits the next? How do they justify their acceptance of such medical technologies as gene therapy or cloning while, at the same time, continuing to deny the same evolutionary principles from which these advances are founded? There is no compromise. One must either accept the doctrines of science–of reason–or one must reject its principles altogether. We either trust in the scientific method or we do not.

One problem many religions have with science is that it represents a source of constant contradiction. For example, in the old days, if the land was dry, men prayed for rain. Since they didn't understand the underlying physical cause of this phenomenon, they believed that the rain's fall was determined by the impulses of those who lived beyond the clouds, by the wills of the gods. How else were humans to explain such a thing? They couldn't. It took humankind thousands of years of scientific discovery and research before we understood the nature of the evaporation and condensation of water molecules–that is, of rain. But we needed some sort of explanation. What else were we to do? Accept that it rained for absolutely no reason whatsoever? This would hardly be possible, as it is human nature to pursue the underlying cause and nature of things.

Today we know better than to believe that rain is produced by the whims of gods. Today, we know that rain occurs because of a series of physical causes and effects. In this way, science has emasculated the old gods. It has stripped them of their powers and has instead allotted them to a source that is wholly neutral, one that is indifferent to the affairs of men, one scientists refer to as "the forces of nature."

Now I can certainly understand why humans would desire to believe in a god, in a force that cares about us, that treats us as its favored creature. Believing in a god provides us with a sense of purpose. It bestows us with immortal life. But should we believe in such things if it's at the expense of everything that corresponds with reason?

And so, at the age of twenty-one, I decided to place my faith in the physical sciences. And why not? At this point, I had every reason to believe in the logic of the physical universe and none whatsoever to believe in any spiritual reality. Until proven otherwise, I would pursue all things, including the nature of God's existence, from a strictly physical–that is, a scientific–perspective.

Only how was one to use science to find God? Into what constellation does one point his telescope? What slide is one to place under the microscope?

...And so, my quest continued.

Chapter 3

A Very Brief History of Time

or

Everything You Ever Wanted to Know about the Universe but Were Afraid to Ask

"To be master of any branch of knowledge, you must master those which lie next to it; and thus to know anything, you must know all."

—Oliver Wendell Holmes

"And I gave my heart to seek and search out by wisdom all things that are done under heaven."

—The Old Testament, Ecclesiastes

"Canst thou by searching find out God?"

—The Old Testament, Book of Job

So off I went, full speed ahead, searching through numerous scientific tomes...for God. There was physics, chemistry, biology, physiology, psychology, geology, astronomy, and cosmology, to name a few, each one a school unto itself.

The more I studied the various sciences, however, the more I realized how much they were all so integrally interrelated. It was as if the scientists had somehow made the mistake of breaking the unified history of the entire physical universe into several separate epochs or categories without recognizing that they were each linked to one another in the most essential way. And so, the more I studied, the more I came to realize that science was simply the study of the history of the entire physical universe from the dawn of time.

As I embarked on my newfound quest for a scientific interpretation of God, I decided to begin with physics, as it seemed to address nature's most fundamental principles. From physics I learned how the universe emerged approximately fourteen billion years ago, at which time all the matter in the universe was condensed into one single, solitary point of pure energy. The pressure within this point was apparently so great that it erupted in an enormous explosion, which, in turn, released all of the universe's energy outward into vast space, an event scientists refer to as the "big bang."

As Einstein taught us, energy and mass (matter) are interchangeable: $E=MC^2$. Energy equals mass times the speed of light (approximately 186,000 miles per second) squared. What this essentially means is that if mass (matter) is accelerated to a high enough speed, it will become energy. Inversely, should energy be slowed down, it will settle into matter. And so, within one-millionth of a second after the universe's initial eruption, energy began to settle into its first material particles. By one ten-thousandth of a second after the big bang, forces inherent within these first infinitesimal particles prompted them to bond with one another to form larger infinitesimal particles. Three minutes after these first "subatomic" particles had formed, they settled into the first stable material objects known as "atoms," lithium, deuterium, and hydrogen atoms, to be exact.

For the first four hundred million years after this initial eruption occurred, the universe existed as an expanding cloud of predominantly hydrogen atoms, which, due to the initial force of the big bang, were being propelled further outward into vast space.

The law of gravity states that all matter is attracted to all other matter. It was this force, inherent within the hydrogen atoms, that prompted them to gravitate toward one another, causing them to congregate into vast gaseous clouds.

Now there were two forces working on the hydrogen atoms simultaneously, one that propelled them outward into space and another causing them to gravitate laterally toward one another. This second force continued to act upon the hydrogen atoms until they had swelled into humongous clouds. Because the force of gravity always falls towards an object's center, the weight of all of this hydrogen collapsing upon itself created a tremendous amount of pressure within these clouds' cores. When the pressure within the cores became more than the hydrogen atoms could withstand, they began to fuse. As a result of this fusion process, four hydrogen atoms are compressed or "fused" together to form a heavier atom we call helium, the next stable form of matter or "element" to exist within the universe. When four hydrogen atoms fuse to create one helium atom, not all of their mass is retained within the helium. Instead, some of the hydrogen's mass is lost as energy radiated outward in the form of heat and light. The moment one of these hydrogen clouds begins this fusion process, we refer to it as a star, our own sun a perfect example.

Millions of years after a typical star is born, after the majority of its hydrogen atoms have already fused, it begins fusing its heavier element, its helium. When helium atoms fuse, they are transmuted into the even heavier element of carbon. As this process continues, newer, heavier atoms or elements are created within a star's core. After a star depletes itself of most of its fusible matter, it becomes unstable, often causing it to erupt in a tremendous explosion called a supernova. As a result of a supernova, all of a star's newfound elements are dispersed throughout the ever-expanding universe.

It was at this point that I noticed my physics texts were coming to a close and that my chemistry books were just beginning. It seemed that once these newly created elements began interacting with one another, the history of the universe had been divided into a whole new field of study, almost as if it had been arbitrarily broken into separate chapters. In finishing "Physics," I had just completed chapter one in this cosmic serial. It was now time to move on to the next installment in the history of the universe–Chapter Two: Chemistry.

Physics had outlined the essential forces of nature, forces inherent in all matter. When dealing with how these forces affected matter's smallest particles, it was referred to as quantum, particle, or atomic physics. When dealing with how these forces affected the interaction of much larger objects such as planets or stars, it was called astronomy. When dealing with the full scope of all the energy and matter that existed within the entire physical universe, it was cosmology.

After physics had left me with an explanation of the various atomic forces as well as how the various elements were formed, physical chemistry sought to explain the dynamic involved in those interactions that occurred between the various atoms. Since each new element created within these fiery stars consisted of a different number of electrons (a subatomic particle carrying a negative charge), each atom carried a slightly different electrical charge from all others. Based on their relative charges, some of the differing atoms began to bond with one another to form more stable particles known as compounds or molecules. Chemistry sought to interpret the unique set of properties that each one of these new atomic combinations contained, as well as how they reacted with one another. One sodium atom and one chlorine atom, for instance, have a propensity to bond with one another, creating a compound we call sodium chloride, more commonly known as salt. With this new diversity of atoms being distributed throughout the universe, an abundance of new molecular combinations began to emerge. From its humble beginnings, when it consisted almost entirely of hydrogen, the universe had evolved into a complex array of physical compositions.

Depending on such variables as pressure or temperature, any compound could exist in one of three forms–a solid, a liquid, or a gas. Many of the compounds, as they existed in solid form, were referred to as minerals. As a result of the attracting nature of electromagnetic and gravitational forces, these minerals began to cluster together into ever-larger formations.

Quick cut to astronomy: nearly five billion years ago, about nine billion years after the initial "big bang," our sun was formed from a tremendous cloud of gas. Although the vast majority of this rotating cloud's mass was made up of hydrogen, it contained many other, heavier elements as well. As the core of this mass of gases consolidated to become a star, some of the heavier elements dispersed around the cloud's periphery began to amalgamate into large mineral clusters.

When one of these peripheral mineral clusters flies too close to a star, it is drawn in by the star's enormous gravitational pull and absorbed into it. If a cluster's momentum exceeds the star's gravitational pull, it will spin off into deep space. In the rare case that the cluster's momentum happens to be at equilibrium with the star's gravitational pull, it gets caught in the star's gravitational field, causing it to travel in an elliptical course around that star. We refer to such a course as an orbit. When a large enough mineral cluster falls into a star's orbit, we call it a planet. We live on Earth, the third planet from our star, the sun.

Sometimes smaller mineral formations become caught in a planet's gravitational field, causing it to fall into the planet's orbit. We call a mineral cluster that orbits a planet a moon. We call a star combined with all of the planets that orbit it a solar system. Our solar system consists of a star (the sun) with nine planets (Mercury, Venus, Earth, Mars, Jupiter, Saturn, Uranus, Neptune, and Pluto) orbiting around it. On an even larger scope, a cluster of solar systems is called a galaxy. All the galaxies in vast space make up the universe.

Meanwhile, back to our star's own spinning satellite, back to planet Earth. Enter the science known as geology. Approximately 4.6 billion years ago, the Earth was formed. At that time, the Earth was little more than an enormous ball of molten rock. Not yet possessing an atmosphere to shield it from falling celestial debris, the Earth was constantly being bombarded by stray mineral clusters known as meteorites. As these meteorites continued to shower the Earth, the planet continued to increase in mass and size.

Moreover, when these meteorites hit the Earth, enormous amounts of heat energy were unleashed with each tremendous impact, reducing them to molten form. As a result, gases that had previously been trapped within the meteorites were released into the Earth's incipient atmosphere.

Since gases are light and volatile, they have a tendency to fly away from a planet and to dissipate into space. A planet like Mercury, for instance, is so small it doesn't have a strong enough gravitational pull to retain such light and volatile particles and therefore has no atmosphere. Some planets, such as Jupiter, are so large that their gravitational pulls cause their gaseous elements to be so firmly drawn to the planet's surface they become condensed into liquid pools, and therefore also lack a viable atmosphere.

The Earth, however, was neither too small to retain its gaseous particles, nor was it so large that it compressed them to its surface. It was neither too close to the sun (the heat of which affects the volatility of these gases) that the gases were propelled off into space, nor was it so far from the sun that they became frozen into solid form. Instead, the conditions on Earth were such that any released gases were held within close enough proximity to the surface that they came to form a gaseous shell around the planet. We call this shell the atmosphere. Once the atmosphere had formed, when a meteorite got caught in the Earth's gravitational pull, the friction incurred by the meteorite rubbing against the atmosphere's gaseous particles caused a falling meteorite to burn up before it could reach the Earth's surface. No longer vulnerable to the heat-emitting collisions generated by falling meteorites, the Earth began to cool.

Two of the gases most often trapped within these falling meteorites were hydrogen and oxygen. Consequently, an enormous amount of these two elements began to fill the Earth's atmosphere. Due to their prospective electrical valences or charges, oxygen and hydrogen began to bond with one another to form a molecule commonly known as water. As water molecules now began to accumulate within the Earth's atmosphere, they began to aggregate into a dense vapor that eventually succumbed to the planet's gravitational pull, causing them to be drawn back down to the Earth's surface in the form of droplets we call rain. When these first rains fell to the Earth, they caused the planet's molten surface to cool even further, in turn, prompting even more trapped gases to be released in the form of steam. More water vapor yielded even more rain, which caused the planet to cool even further.

This process continued for nearly a billion years, after which approximately two-thirds of the Earth had become covered in water with the other third made up of a hardened mineral shell. Within these oceans of water, there stirred a broth consisting of ammonia, methane, water, sulfur dioxide, and hydrogen.

In 1953, a researcher by the name of Stanley Miller put this information to use by conducting a very important experiment:

> Miller set up an airtight apparatus in which the four [original primordial] gases could be circulated past electrical discharges from tungsten electrodes [patterned after the primordial Earth's lightning storms]. He kept the gases circulating continuously in this way for one week, and then analyzed the contents of his apparatus. He found that an amazing number and variety of organic compounds had been synthesized. Among these were some of the biologically most important amino acids as well as such substances as urea, hydrogen cyanide, [and] acetic and lactic acid.[1]

Within the confines of his laboratory, Miller had simulated the Earth's chemical evolution. He had synthesized amino acids, the building blocks of all organic matter, the essence of all life. In doing so, Miller had accomplished what was formerly believed to be the exclusive privilege of gods. And yet, here it was, organic evolution without God...just Stanley Miller with his airtight vessel of chemicals, a flame, and a little electricity.

Starting with a composition consisting almost entirely of hydrogen, the universe had evolved, almost ten billion years after its conception, to a point in which it contained complex chains of macromolecules. Macromolecules that contained carbon possessed such unique properties that my chemistry books had suddenly diverged into a whole new science called organic or biochemistry. I now had to purchase a whole new set of texts that dealt exclusively with these complex carbon-based compounds, ones similar to those Miller had synthesized in his lab.

Back to Earth: For the next billion years, these complex organic (carbon-based) compounds brewed and churned within the Earth's primordial seas, within which trillions of various molecular combinations emerged, each possessing a unique set of physical and chemical properties. Many of these molecular combinations to emerge were so complex that inherent instabilities caused them to eventually disintegrate back into their contingent parts.

As these larger and more complex molecules continued to brew in the Earth's seas, new combinations were constantly being forged, each one slightly different from the next. Among these "organic" molecules, some of the variations to emerge possessed the capacity to absorb the Earth's and the sun's radiated heat and light energies. With this newfound capacity, otherwise unstable molecules were now able to use these external energy sources as a means to maintain stability.

Even with this new capacity, none of these energy-absorbing macromolecules were efficient enough to overcome their inherent instabilities altogether. Being able to harness the sun's energy merely allowed these complex molecular chains to maintain their structural integrity for a slightly longer duration. Even so, it was still just a matter of time before these molecules succumbed to inherent instabilities and eventually disintegrated back to their contingent parts.

As newer variations of these energy-absorbing, carbon-based macromolecules continued to churn within the Earth's primordial seas, some eventually emerged with a newfound capacity to produce duplicates of themselves before they disintegrated. These new molecules could now ensure the preservation of their physical identities through the continued existence of their duplicates. Due to the disruptive effects of the sun's ultraviolet rays, however, not all of these duplicates turned out to be identical to the "parent" molecule from which they came. Among these slight variations to arise, most were harmful and worked against the preservation of the "daughter" molecule. Nevertheless, some of these variations happened to be even more energy-efficient than their parent molecules, in which case the new design would often supersede the old one. As this process continued, more energy-efficient molecular combinations emerged.

In time, these complex carbon-based macromolecules evolved other capacities that maximized their potentials to maintain stability. Some of the other capacities these macromolecules had evolved included ingestion (the capacity to absorb energy), digestion (the capacity to assimilate ingested energy), excretion (the capacity of the macromolecule to rid itself of any of its digested energy's harmful by-products), and locomotion (the capacity to move from one place or position to another). As these self-replicating, energy-absorbing macromolecules continued to evolve, I noticed that my organic chemistry books were also evolving into a new science called biology.

As with all the other sciences, biology came with its own terminology. In biology, for example, molecules that could perform the aforementioned functions were now referred to as "living." When a molecule made a copy of itself, this was now referred to as "birth." When, in time, one of these molecules eventually disintegrated, it was now called "death."

The first forms of life to exist reproduced asexually, meaning they required only one parent cell that would divide into two separate daughter cells. Once again, due to the disruptive effects of the sun's radiation, many of these offspring contained slight mutations that made them vary, to some small degree, from their predecessor's design. Variations that were more energy-efficient were more likely to survive. Those most likely to survive were most likely to duplicate themselves and therefore to pass along their advantageous characteristics (traits). On the other hand, those variations that were least energy-efficient were most likely to be discontinued. My biology books had a very specific term for this organic weeding process: natural selection. As a result of this process of natural selection, organic matter–life–continued to evolve.

In order to keep inventory of these constantly diverging "living" compositions of matter, biologists classified them into various categories based on their inherent characteristics. The first varieties of life to emerge on Earth diverged into two distinct branches. One used the Earth's oxygen to establish its energy supply, while the other used carbon dioxide. Biologists divided these first two living forms into two separate classifications known as kingdoms. Those forms that used carbon dioxide to supplement their fuel supply were classified as belonging to the plant kingdom, while those which used oxygen were categorized as belonging to the animal kingdom. As time passed, these two kingdoms continued to diversify, each producing a vast array of unique forms (species). Within the next three billion years, a myriad of these species propagated across the planet, blanketing the Earth's surface with a thin organic shell.

Three billion years after life had first evolved, the seas were suffused with a variety of these plant and animal forms. It was at about this time that one of these sea-dwelling animals evolved a spinal cord, a protective sheath that enveloped the organism's nervous system and helped to distribute its nerve cells throughout the length of its body. This represented the beginning of a new classification of animals biologists referred to as the subphylum vertebrate. As the vertebrates continued to diverge, biologists placed them into separate categories known as "classes." The first class of vertebrates to emerge were the fish.

About a hundred million years later, some of these fish evolved the capacity to survive on land as well as in the water. These biologists classified as amphibians. About a hundred million years after that, a newer class of vertebrates evolved from the amphibians, one which lived exclusively on land. These were called reptiles.

Within the next fifty million years, some of the reptiles evolved in such a way that their scales were replaced by feathers, their bones became hollow and they developed the capacity for flight. These were the birds. Approximately another forty million years after that, yet another land-dwelling creature emerged from the reptiles. These were the mammals. Mammals were different from their ancestors, the reptiles, in that their embryos developed from within the mother's body rather than from within an externally incubated egg. Mammals produced milk with which they could feed their young. They were coated with hair, homeothermic (warm-blooded), and, most significantly, developed a much larger brain that allowed them to respond to their environments in a much more sophisticated manner than all the Earth's other living forms.

Among the mammals, sixteen subclasses known as orders emerged. Examples of some of these orders were rodentia (rats, mice, squirrels, etc.), carnivores (cats, dogs, bears, etc.), cetaceans (dolphins, whales, porpoises), and artiodactyla (cattle, sheep, goats, deer, etc.). About a hundred million years after the mammals first evolved, approximately fifty million years ago, a particular mammalian order emerged, known as the primates. Primates differed from the other mammals in that they evolved such adaptive features as stereoscopic vision, enhanced mobility of the digits (fingers) complemented by an opposable thumb, and larger brains–particularly a larger cerebral cortex (that portion of the brain where memories are stored and most cognitive processing takes place).

As time went on, these primates continued to diversify until they had evolved into a family called the hominids. Hominids stood upright, as compared to their ancestors that walked on all fours. With the advent of this new adaptation, these animals now had two free limbs with which they could hold, carry, and manipulate objects at the same time that they could transport themselves. The hominids continued to evolve until about a hundred thousand years ago when they reached their apex with the emergence of a new species known as Homo sapiens, more commonly known as humans. This human animal had evolved vocal cords with which it could enunciate a variety of sounds, thus enhancing its capacity to communicate with others. Furthermore, humans evolved certain structures within their brains that allowed them to organize these sounds in such a way that they could create and speak words–combinations of sounds that symbolized objects. The use of words enabled humans to communicate ideas with advanced precision. Such qualities as these combined with an enhanced capacity to store and process information made Homo sapiens Earth's most powerful creature.

Before I delve any further into the subsequent disciplines that pertain exclusively to the human animal, I would like to clarify a few things. In a matter of pages, I have jumped from the origin of the first organic matter to the emergence of humankind. But by what process does such an evolution take place? How is it possible that within three and a half billion years, a simple cell membrane could have turned into flesh, a vacuole into a complex digestive system, a cellular nucleus into a brain? How could a reptile's scales become feathers or its legs become wings? What kind of organic alchemy or molecular witchcraft was this that could transform creatures from one thing into another? To offer an illustration, let's take the example of a human being.

Two cells, a sperm and an egg, meet. These two cells happen to be distinct from all others within the human body in that each carries only half of its host's chromosomes. Within the sperm cell's nucleus lie half of the father's chromosomes, within the egg cell, half of the mother's. When these two chromosomally incomplete cells meet, when the egg becomes fertilized, the two sets of chromosomes merge and recombine to form one unique and chromosomally complete cell.

This now complete set of chromosomes within the newly fertilized cell is like a blueprint that contains all the material required to create a fully developed human being. The chromosomes themselves are composed of sections called genes. Each gene contains information to create one or more of what will soon unfold to become that individual's physical traits. For instance, whereas one gene might carry information that will determine a person's sex, another might carry information that will determine skin color, another that person's height, hair color, etc. This list of physical features goes on until one's entire anatomy, from the shape of one's head to the soles of one's feet, has been accounted for–all of it stored within the contents of one's genes.

But what are genes? According to the biologist William Keeton, a gene is a "unit of inheritance; a portion of a DNA [Deoxyribose nucleic acid] molecule."[2] Here is Keeton's technical description of this molecule:

> The molecule has a ladderlike structure, with the two uprights composed of alternating sugar and phosphate groups and the cross rungs composed of paired nitrogenous bases. Each cross rung has one purine base (any one of several double-ringed nitrogenous bases) and one pyrimidine (any one of several single-ringed nitrogenous bases). When the purine is guanine, then the pyrimidine with which it is paired is always cytosine; when the purine is adenine then the pyrimidine is thymine. Adenine and thymine are linked by two hydrogen bonds, guanine and cytosine by three.[3]

So, genes are made of DNA, a macromolecule consisting of a combination of sugar molecules, phosphate molecules, and nitrogen-based molecules, all ordered into a twisted, ladderlike structure known as a double helix. In essence, genes are made up of molecules. And what are molecules? Molecules are arrays of two or more atoms. For instance, a sugar molecule, like the one in DNA, is made up of a combination of carbon, oxygen, and hydrogen atoms.

Carbon, oxygen, and hydrogen atoms; nitrogenous bases; phosphates: these are the essential ingredients needed in the recipe for making a human being. Stored in the particular arrangement of these atoms exists all the information necessary to create a person's entire physical makeup, all accounted for before that person is even a fully emerged embryo, let alone born. A person's sex, skin and eye color, height, vision, hearing, and proclivity for such mental or physical diseases as asthma, diabetes, schizophrenia, Alzheimer's, and allergies, as well as such personality components as propensities for shyness, aggression, curiosity, depression, athleticism, musicality, math ability, joviality, as a mere few examples, all existing within this first fertilized cell–the essence of our physical and psychological life story told from the very first moment we are conceived.

So the sperm and the egg meet to create one very informed fertilized cell. Stored within this first cell are instructions to divide. Once this occurs, the emerging person exists as two cells, each now containing all the information necessary to create a fully developed human being. These two cells will now reproduce, and so on and so on, until a cluster of cells are formed. Stored within each of these cell's chromosomes is information that will now instruct the cells to start producing even more specialized ones, such as nerve cells, blood cells, and muscle cells. With the emergence of these specialized cells, the unborn embryo will continue to differentiate and grow within the mother's womb until nine months later when it is ready to be born.

So all of our traits are, generally speaking, predetermined from the moment of our conception. But what exactly are traits? Traits are those characteristics that distinguish not only one species from the next but each individual within a species from every other. And from where do these traits emerge? They originate from information stored within an organism's genes, that unique arrangement of atoms that make up an organism's chromosomes.

For example, the fact that all fish have gills would imply that somewhere in a fish's chromosomes lies a gene or group of genes that instructs the developing fish embryo to produce gills. This is not just true of the fish's gills but of every single physical characteristic a fish possesses. As no trait can develop of its own volition, this means that for every trait fish possess, there must exist some corresponding gene or group of genes responsible for its emergence. Unless we are to believe that all fish have gills as the result of some incredible accident or coincidence, we must accept the genetic, evolutionary explanation for such a phenomenon. If fish possess gills, there must exist "gill" genes. If a fish is equipped with fins, there must exist "fin" genes, and so on and so on, until every single physiological characteristic of a fish is accounted for. In this way, the developed animal is a composite of traits that correspond to information stored within an animal's genes, once again, information already established from the moment that animal is conceived.

As each species possesses its own unique set of traits, each species must possess its own unique set of genes. The fact that fish possess gills means that the molecular arrangement of their genes must be different from a creature that has no gills. The fact that all fish possess gills (excluding, of course, those extreme mutations which represent exceptions to the rule) means that gill genes must exist in all fish DNA.

Since each individual that emerges from a sexually reproducing organism is formed from a unique admixture of its two parents' chromosomes, each individual varies to some degree from every other. In this way, though all fish may possess gill genes, each fish's gills will in some slight way vary from one individual fish to the next.

The same is true for humans. Though we all possess genes that instruct our bodies to develop two eyes, each person's eyes are slightly different. This is true for every characteristic we possess as a species. Whether we are discussing one's height, sense of hearing, skeletal or facial structure, the constitution of one's heart, kidneys, or immune system, each part of us varies in some way from one individual to the next. In a sense, every single part of us, from every cell to every organ, is as unique to each individual as are one's fingerprints, which, though we all possess them, no two are exactly alike.

Regarding these slight variations between individuals, in the constant competition for life, those creatures whose variations are best suited or adapted to their surroundings are at a considerable advantage and are therefore more likely to survive. Those forms more likely to survive will, in turn, have a greater chance of reproducing. Those that have a greater chance of reproducing will, consequently, have a greater chance of passing their genes, along with their advantageous traits, on to future generations.

Just as no two individuals are ever alike, neither is the gene pool of any two generations of a given species. Because each generation is put through another screening of natural selection, each generation will most probably be better suited to its environment. In this way, life is in a state of constant flux, each species constantly maturing and evolving with each passing generation.

Let me provide a hypothetical illustration of how this process of natural selection works: Imagine a place where the land is flat, lush with plants and trees. Roaming this land is a hypothetical three foot tall, horselike creature I will call the nequus. A male nequus and a female nequus mate and have three baby nequuses. Given the way the two parents' genes recombine, it's inevitable that the three offspring will be different from one another. Regarding, for example, the offspring's heights, based on the laws of genetic variance, it's possible that any of the three will end up either shorter or taller than its parents. Back to the nequus plains: Imagine a geological event were to occur that now transforms this once lush region into an arid one. Amid these new environmental parameters, much of the plant life has died. The nequuses, which are herbivores, suddenly find themselves in fierce competition for what remains of their now dwindling food supply. Unfortunately, the average nequus, which is only about three feet tall, can only reach the bottom branches of its region's trees, much of which have already been eaten.

Back to our offspring: Because it can reach the leaves of those higher branches that the majority of its starving species cannot, the tallest of the three is most likely to live long enough to reproduce and therefore to pass its genes onto future generations.

Let's now imagine that this taller nequus, unlike its shorter siblings, which are less likely to survive, lives long enough to mate, thereby passing its "taller" genes onto its offspring. As was true for the father, the tallest of this newest litter is also most likely to survive, thereby passing its "taller" genes onto its offspring. As this dynamic is repeated over a period of multiple generations, it's quite likely that the average height of the nequus will now be taller than its predecessors. In this way, every species is in a state of constant flux, incessantly being modified to most effectively meet the demands of its ever-changing physical environment. Sometimes these evolutionary fluctuations occur in a slow and steady progression that transforms species over a protracted period of time. Other times, a beneficial genetic mutation emerges that is so dramatically different from its peers that a species can be transformed within a few generations (this revision of basic Darwinism was originally postulated by Stephen J. Gould in a theory he called *punctuated equilibria* which purports that the creation of new species sometimes occurs in rapid spurts–rather than in slow progression–which are then followed by long periods of stability).

In the case of the imaginary nequus, should the drought and consequent food shortage continue, the forces of natural selection will continue to weed out those least equipped to survive these conditions and to preserve those that are best. Perhaps after a period of ten million years of such natural selection (what would amount to the passing of approximately a hundred thousand generations), the average height of a nequus may have grown to be ten feet tall, making it resemble something more like a giraffe than a horse. In essence, what used to be a nequus has now evolved into a different species with a new sequence of genes. Apparently, necessity is the mother of selection.

To provide an actual example of how environmental pressure can alter a species' physiology, I'll now refer to the real-life case of the Biston betularia or what is more commonly known as the peppered moth. During the 1800s, it was noticed that this once predominantly white mottled moth had, within a very short period of time, evolved into a much darker variety. Originally, the lighter variety had spent much of its time resting on trees whose bark matched their wings' pigmentation, thus making it much more difficult for predatory animals to see them, an adaptive mechanism known as camouflage. With the advent of the industrial revolution, however, residue from nearby factories covered the forests with dirt and soot, darkening the surface of the trees. Because the white moths, which represented the majority of the species, could now be more easily sighted by predators, they became more likely to be eaten. In contrast, the darker variety of the moth's population, which previously represented a small minority, were now less likely to be seen by predators and, consequently, that much less likely to be eaten. Because they were less likely to be eaten, the darker variety were now more likely to survive long enough to pass their genes on to future generations. As a result of this sudden change in the environment, the moth's population had quickly shifted so that the species' darker strain, once the minority, now came to represent its majority. And so, within just a few generations, the entire peppered moth population had been modified due to a change in the animal's environment.

Another aspect underlying the forces of evolution involves a process known as genetic drift. To illustrate this process, imagine that due to overpopulation, certain members of a species find themselves having to migrate to a new area in search of new food supplies. For instance, ten finches among a community of tens of thousands migrate to a nearby island in search of food. Since these ten finches can never represent the exact genetic mean of their species, should they reproduce, they will be creating an entirely different genetic pool based on their own particular genetic makeups. In a sense, these ten "pioneer" finches would represent the founders of a whole new, slightly different genetic strain. Because of the pioneer group's slight genetic variance from the mean of its original population, this new strain might, in time, come to represent a whole new species. As a matter of fact, this is exactly what Charles Darwin discovered when he went to the Galapagos to study the various finch species as they existed on each of the archipelago's separate islands. Through his observations, Darwin noticed that the finches from each of the Galapagos' islands seemed to constitute their own unique subspecies. It was from these observations that Darwin first conceived of his theory of evolution.

Returning to the study of Homo sapiens: With the advent of humans, there came a whole new panoply of specifically human sciences, the first of these being anthropology. Anthropology dealt with matters concerning the social, behavioral, and physical evolution of those advanced primates, the hominids, all the way up until about ten thousand years ago when humans reached what is referred to as the Neolithic stage of their existence. What separates Neolithic humans from their biologically identical ancestors was the discovery of agriculture. Before the Neolithic period (during what is known as Man's Paleolithic age), these more primitive humans wandered the globe in nomadic tribes, constantly moving from place to place in search of new food supplies.

But humans possessed an evolved brain and, over time, began to notice patterns in their world. Unlike any other animal that came before them, humans could recognize, for example, that where a plant's seed had fallen, a new plant would often emerge. When the first humans made this connection, about twelve thousand years ago, it enabled them to imitate nature by planting their own crops. With the advent of agriculture, the human animal began settling into stationary communities (usually near a river which allowed for a constant water supply as well as a means of transportation). Furthermore, by noting the manner in which other animals reproduced, humans learned to herd these animals so as to control their meat supply to supplement their diet of fruits and vegetables. The combination of these two events is referred to as the agricultural revolution. It is referred to as a revolution because of the immense impact these discoveries had on our species. For the first time in our species' history, humans could regulate their own food supply. No longer needing to devote all of their time to searching for their next meal, humans could afford themselves some extra or what we call leisure time. With all this additional time on their hands, human societies now had the opportunity to direct their energies to self-expression (the arts), play (sports), as well as the pursuit of wisdom and knowledge (philosophy and science).

As some of these agricultural settlements began to flourish, other peoples began to migrate to them hoping to reap the benefits of these new establishments. In time, these settlements began to expand in size and population. It was here, in these first cities, where humans from a variety of cultures first congregated in order to exchange goods as well as ideas. This marked the dawn of a period in our species' history known as the urban revolution. As these cities continued to grow, humankind's first civilizations arose.

As time went on, civilizations rose and fell. Without reciting the histories of all the various civilizations, suffice to say that this process continued until we find ourselves here today at the dawn of the twenty-first century.

Now I make no claim that science could explain everything. Sure, there were parts of the physical universe that were better understood than others. Sure, there were whole fields that were, in many ways, still incipient and, consequently, theoretical in nature. Sure, there were still mistakes to be made, details to be reworked and revised. Generally speaking, however, the scientific interpretation of the universe always remained true to its method, one that has given us nuclear energy, organ transplants, electric light, and antibiotics, as a mere few examples of its awesome capacity. Here were technologies that I knew as a fact worked. These things took a great deal of scientific research to create: the exact same type of research and methodology that was used to account for the aforementioned history of the entire physical universe. Essentially, the proof was in the products. If I could rely on the scientific method to create such wonders as space shuttles, gene therapies, nuclear power, and microwave ovens, then why shouldn't that same methodology be able to explain the origin and evolution of the entire physical universe as well as of all terrestrial life? How else could science have so successfully mastered and manipulated our physical world if it didn't understand its very nature?

Science had accounted for the approximately fourteen-billion-year history of the entire physical universe from its origins to its present state and all without the aid or assistance of any spiritual entity: *Cosmology without God!* Science had been equally able to account for the approximately three and a half billion years of organic evolution, also without the aid or assistance of any transcendental force or being: *The origin and evolution of life without God!* No longer was either life or the universe contingent upon the existence of some intervening deity. Not to say this meant that God didn't exist, but let's just say it bolstered the possibility.

No longer would I have to ask such questions as, "If there is no God, then how is one to explain the origin of life?" Or "Without God, how did the Earth, the Moon, the Sun, and the stars all come to exist?" No longer would I have to look down at my own body and not understand the origin, evolution, nature, and mechanics of my own being.

All this, science had done for me. First it rescued me from the clutches of mental illness, and now it had made the universe comprehensible to me. And yet, there it was, taunting me as much as ever–that incessant longing, that gnawing need to know not *how* I or the rest of the universe worked but *why*? There it was still looming over me, as oppressive as ever, that relentless problem of the meaning of my existence. Why was I here? What was my purpose? As always, underlying this question was the elusive problem of God. Only knowledge of God could resolve the ultimate question of my existence. And yet, how was it that amid all of this glorious information the sciences had yielded, it couldn't offer me any explanation whatsoever regarding the nature of God's existence? Was God simply incomprehensible to us? Or was there a scientific explanation, only no one had discovered it yet? What pattern in nature, what empirical observation, I wondered, might possibly help to reveal the nature of God's existence to humankind? Then again, even if there was a solution, might it lie beyond our reach, a problem meant to torment and tantalize us until the end of time?

Regardless of whether the problem was answerable or not, all I knew was that, spiritually speaking, I had yet to be satisfied. The quest would have to go on.

CHAPTER 4

KANT

"What is real? How do you define real? If you're talking about what you can feel, what you can taste, what you can smell and see, then real is simply electrical signals being interpreted by your brain."

–*THE MATRIX*

"All that I experience is psychic. Even physical pain is a psychic event that belongs to my experience. My sense-impressions–for all that they force upon me a world of impenetrable objects occupying space–are psychic images and these alone are the immediate objects of my consciousness. My own psyche even transforms and falsifies reality, and it does this to such a degree that I must resort to artificial means to determine

> what things are like apart from myself. Then I discover that a tone is a vibration of air of such and such a frequency, or that a color is a wavelength of light of such and such a length. We are in all truth so enclosed by psychic images that we cannot penetrate to the essence of things external to ourselves. All our knowledge is conditioned by the psyche which, because it alone is immediate, is superlatively real. Here there is a reality to which the psychologist can appeal, namely psychic reality."
>
> –CARL JUNG

So far, my search for knowledge of God had been directed outward onto those objects that constituted the entire physical universe. I had studied the physical nature of atoms and molecules, of planets and stars, of organic and inorganic compositions of matter. And still, no matter where the astronomers had pointed their telescopes, or what specimens the biologists had placed under their microscopes, or which particles the atomic physicists had split asunder, not one had ascertained anything resembling verifiable knowledge of any spiritual reality or God. And so, in order to complement my investigation into the physical sciences, I was simultaneously studying the often enigmatic discipline known as philosophy.

Though its Greek roots translate to mean "love of wisdom," philosophy, as I saw it, constituted the study of the ultimate nature of reality. What, if anything, can be said to be real? What, if anything, can be said to represent truth? In essence, what is reality?

The ancient Greeks, who are generally recognized as the founders of Western philosophical thought, believed that in order to understand the ultimate nature of reality one had to first understand the nature of all things that encompassed the vast physical universe. What, for instance, are the various things that make up our world made of? Where did they come from? In what ways are they similar? In what ways are they different? These were the types of questions the ancient Greeks felt needed to be resolved if the true nature of reality was ever to be fathomed.

Similar to the Greek method, this was how I, too, had been conducting my own personal investigation–by studying the nature of those material objects that permeated the fourteen-billion year history of the entire physical universe. This was the method by which I too sought to comprehend the nature of ultimate reality, a problem I presumed would, once resolved, lead me to an even more comprehensive knowledge of spirit and God. I, like the Greeks, had been looking outward for my answers–that is, until I came upon the work of the eighteenth-century German philosopher, Immanuel Kant.

Since the ancient Greeks first introduced this particular method of inquiry (of looking into the nature of things external to them), this represented the predominant trend in all human science and philosophy up until the eighteenth century when Immanuel Kant arrived on the scene. In his book *Critique of Pure Reason*, Kant had made one of the most revolutionary leaps in the history of human thought by suggesting that in order to understand the true nature of reality, we need to redirect the focus of our inquiries from outwards to within. Kant proposed we do this by studying not the nature of those physical objects around us but rather the manner in which we perceive those objects. Rather than looking outward for answers regarding the ultimate nature of reality, we first need to look inward, into the nature of that which is doing the perceiving, into the nature of perception itself.

Take, for example, an apple. By what means, Kant asked, do we come to have knowledge of an apple? The answer: through information we acquire through our physical sense organs. Through the absorption of reflected photons of light as they fall on our retinas and are then processed by our optic nerve, we see the apple. Through molecules the apple releases into the air, which are then picked up by the brain's olfactory, we smell it. As its chemistry dissolves on our tongues, triggering electric currents sent to the brain, we taste it. Through the pressure of the apple's contours against our skin, inciting electrical signals up our arms and to the brain, we feel it. Only after our brain processes this medley of electrochemical information do we come to possess knowledge of the apple as an integrated object. What we therefore call an apple is, in actuality, nothing more than electrical signals as they are interpreted by our brain. By this reasoning, it isn't the apple *itself* we come to "know" but merely the apple *as we perceive it*, that is, as our brains filter and process it. Consequently, we can only "know" what our brains enable us.

Due to this perceptual limitation, Kant asserted that it is not within the realm of possibility for us to ever possess absolute knowledge of any object or thing. Such absolute or objective knowledge Kant referred to as noumena–the unknowable world of "things in themselves." Instead, Kant posited, we can only possess subjective knowledge of "things as we perceive them," what he referred to as phenomena. Consequently, all that we call knowledge is relative to the manner in which we perceive and therefore interpret reality.

Kant's ideas actually represented the evolution of the thoughts of the seventeenth-century English philosopher John Locke. According to Locke, humans are born as clean slates, a "tabula rasa" as he phrased it, our minds "empty tablets capable of receiving all sorts of imprints but have none stamped on them by nature."

Almost a hundred years later, Kant wondered: how was it possible that the multitude of data that incessantly passes through our sense organs could so spontaneously arrange itself in such a way as to yield coherent information to us? How is it that this vast stream of stimuli we are constantly being bombarded with manages to fall into place in such an intelligible manner? According to Locke, this process occurs automatically. Not so, Kant contended.

According to Kant, there was just no way this multitude of sense-impressions could arrange itself in such an effective manner of its own volition. Apparently the human mind, Kant contended, must be born, not as a clean slate, but with built-in "modes of perception" that work to organize the multitude of information our sense organs are constantly imparting to us. Without such built-in processing mechanisms, we would experience reality as an unintelligible jumble of sense-experiences. It is therefore necessary, Kant argued, that there exist inherent structures of the mind that function to order the profusion of sense-experiences we receive. The human mind is therefore not some passive organ, as Locke would have had us believe, waiting for experience alone to shape and define us, but rather an active one that works to bring order to the vast array of information we receive.

Two of the many ways that Kant speculated humans inherently process information were temporally and spatially. According to Kant, humans are equipped with built-in processing mechanisms that work to provide spatial and temporal order to our experiences. Accordingly, space and time are therefore not things we perceive "in themselves" but rather represent two innate modes of perception–what Kant referred to as "categories of understanding"–through which our species processes all information. Our comprehensions of time and space are therefore not concepts we learn through experience but rather represent two of the means through which we inherently perceive and, consequently, interpret reality.

As I contemplated Kant's ideas, I recalled the work of the developmental psychologist Jean Piaget. Based on a series of experiments he conducted, Piaget concluded that children can only distinguish proportions of time and space after reaching a certain stage in their cognitive developments, one he referred to as "the stage of concrete operations." Piaget found that before this phase in our mental developments, not only are children unable to distinguish proportions of time and space, but they cannot even be taught to comprehend such concepts.

To demonstrate this, Piaget placed two glass beakers before a number of children of different ages. Though one of the beakers was short and wide and the other tall and slender, both were equal in volume. When asked which of the two beakers would hold more liquid, it was the children's inclination to believe that the answer was the tall, narrow one. In order to show that the two beakers were equal in volume, Piaget filled the short, wide one with water. He then poured the contents of this first beaker into the tall, narrow one. Because the two beakers were equal in volume, as the short, wide one emptied, the tall, slender one became filled. What this should have clearly demonstrated was that both containers were equal in volume.

After the demonstration was complete, Piaget again asked the children which container held more liquid. On this second questioning, children ages seven and up almost invariably answered that the two beakers were equal, while those who were younger still believed that the tall, narrow one could hold more. What this showed was that not until children reach a certain age can they even be taught to comprehend certain spatial relationships.

Based on this data (combined with results of similar experiments that dealt with temporal development), Piaget theorized that there exist innate modes of comprehension that guide the means by which we understand and interpret reality. The fact that Piaget had demonstrated that our ability to distinguish temporal and spatial relationships emerges in all humans at approximately the same age suggests that such aptitudes represent an inherent part of our species' natural cognitive development, something Kant had first conceived nearly two centuries earlier.

So perhaps Kant was right. Perhaps humans are born with specific "modes of perception," a variety of ways that the brain innately processes information, ways that ultimately determine the manner in which we, as individuals as well as a species, interpret reality. Was it possible, I now wondered, that I might somehow be able to apply Kant's principles to the subject of human spirituality, that is, to my own personal quest for knowledge of God?

Had I been wasting my energies trying to fathom God's nature by studying those objects that make up the vast physical universe when, instead, I should have been studying the nature of perception? Was it possible that the manner in which we comprehend God was linked to one of our species' inherent modes of perception? Perhaps I needed, as Kant had done, to invert the nature of my quest from outwards to within. Perhaps the solution to the problem of God lay not "out there" but rather somewhere within the workings of my mind or, as my biopsychology texts would have it, the workings of my organ, the brain.

CHAPTER 5

GOD AS WORD

"In the beginning was the word, and the word was with God, and the word was God."

–THE NEW TESTAMENT; JOHN 1:1

I was now thirty-one years old; approximately ten years had passed since I had begun my formal exploration into the natural sciences in the hope that it might yield some small knowledge of spirit or God. In this time, I had learned about the fourteen-billion-year history and evolution of the physical universe. I learned how the universe was born and of its consequent expansion; how the force of gravity would one day overwhelm the universe's expansionary thrust, causing all of its matter to once again collapse upon itself, thereby reuniting all matter and energy into one condensed single point, the same as it was the moment before the last "big bang" occurred; how, at this time, yet another explosion would occur that would cause the whole process to start all over again; how this process of expansion, equilibrium, and contraction; expansion, equilibrium, and contraction would repeat itself over and over again, ad infinitum, like a great pulse in space that would beat until the end of time.

restrict my aimless quest to this one particular fantastical being? Why this obsession with the entity we call God? It was as if the need to comprehend an absolute being was somehow instilled in me. Just as I was driven to seek food, shelter, security, and love in my life, I was driven to possess spiritual certainty, driven to search for knowledge of God. But why? There must have been some reason for this compulsion. Nothing springs from nothing. As science had taught me, everything that occurs in the physical universe has its physical causes. There had to be some reason, some tangible explanation for why this particular obsession persisted in me the way it did.

Perhaps I was insane. How else was I to explain such an abstract compulsion? Only if I were insane, then so was almost everybody else on this planet, for this was not my own personal idiosyncrasy but one that I, oddly enough, shared with nearly every person from every culture I had ever experienced, heard of, or read about, dating as far back as to the origin of my species. What kind of bizarre coincidence was this? Sure, everybody has his or her own eccentricities, but why was it that we all shared this particular one?

Some people explain human behavior as the sum of one's life experiences. Yet, even with all of our unique lives, how was it that every culture from the beginning of our species has maintained a belief in a spiritual/transcendental force or being, a god? How was it that people from every walk of life, every culture, race, age, sex, and class, shared a belief in some form of a spiritual reality? How odd that if I were to sit down with another person of any culture, race, age, sex, or class, I would be able to hold a conversation (providing, of course, that we spoke the same language) concerning the nature of a God or gods, the concept of a soul, and the possibilities of an afterlife. Perhaps this was proof in itself that God existed. What else could explain the fact that billions of people from every generation, from every culture, even isolated ones, had all pondered these very same notions? Unless this was the result of some vast and incredible coincidence, some internal force or instinct must have been responsible for this most peculiar human phenomenon.

And so, I stepped back and typed out the question on my computer screen: "What, if anything, can I say that I know with near certainty about God?" As I pondered my own question, shaking my head in the usual frustration, suddenly, in one radiant and Archimedic moment, it dawned on me. As plain as the nose on my face lay the one small but certain fact for which I had been searching. There it was, spelled out on the computer screen before me; simply, God was a word! As insignificant as it may have seemed, this one word represented the first empirically verifiable thing I could say I knew, as a fact, about God. I could read it, write it, hear myself say it. In Braille, I could even touch it. No doubt about it: God, I could say with empirical certainty, was a word.

Should I doubt my own sanity, I could always look up the word *God* in any dictionary. If this wasn't enough, I could go anywhere in the world and ask those around me if they were familiar with the concept of a supreme spiritual being, a god. Who could deny that at some point in his or her life he or she hasn't at least considered the existence of some spiritual element in the universe? What functional adult has not, at some point, contemplated the concept of a transcendental force or being? Not even an atheist could make such a claim.

So God was a word, a word that represented the concept of a transcendental/spiritual force or being. Even more compelling, here was a concept for which every culture from the beginning of my species, no matter how isolated, possessed their own symbol or word.

And what exactly did science have to say about words? Where, for instance, did they originate? One place: the human mind. Only "mind" seemed an ambiguous term. In nearly all of the religious/philosophical texts I had ever read, allusions were constantly being made to a mind/body dichotomy, implying the two were separate entities, two distinct agents. Mind intimated that consciousness possessed some transcendental quality. It allowed for the existence of a spiritual component in us. As science had never confirmed the existence of such a component, from hereon, I would only use the word "brain." Minds, science could not verify; brains, it could. No differently than we would view a heart or kidney, brains were 100 percent spirit free, purely physical/organic/mechanical in nature.

So God was a word that, like all words, originated from within the workings of the human brain. Before humans existed, there were no such thing as words. Words originated, as did the concept of God, with our species. Now if brains were strictly biological in nature and the word "God" originated from within that same organ, then perhaps the concept of God was somehow inextricably linked to our biological natures as well. Could it be that the concept of God was somehow a product of my species' inherent cognitive processing, the manifestation of an inherent "spiritual" mode of perception? Was it possible that the solution to the problem of God's existence lay not "out there" but rather buried somewhere within the recesses of the human brain?

The one thing I could now say of God with any empirical certainty was that God was a word, which, like all words, was generated from within the human brain. This meant the only fact I now possessed regarding the nature of God's existence came not from something I had perceived from beyond, from "out there," but rather from something that had been generated from within, more specifically, from within the workings of my physical organ, the brain–and not just my brain but from the brains of almost every single person from every single culture dating back to the dawn of my species.

Trying to decide where to best take this notion, I remembered the position held by the sciences that if a behavior was universal to any given species (or, in the case of humans, to all cultures), it most likely represents an inherent characteristic of that species, that is, a genetically inherited trait. And as surely as all human cultures have spoken a language or engaged in sexual reproduction, all cultures have practiced religion in conjunction with a belief in some form of a spiritual reality. Did this then mean that our perceptions of a spiritual realm–of a God–might also represent the consequence of a genetically inherited trait? And if so, how might I possibly prove such a thing?

Chapter 6

Universal Behavioral Patterns

"It is universally acknowledged that there is a great uniformity among the actions of men, in all nations and ages, and that human nature remains still the same in its principles and operations. The same motives always produce the same actions."

–David Hume

"I will analyze the actions and appetites of men as if it were a question of lines, planes, and solids."

–Baruch Spinoza

"One needs to look near at hand to study men, but to study man one must look from afar."

–Jean Jacques Rousseau

Any physical characteristic that is universally shared by every individual of a given species most probably represents a genetically inherited trait. For example, the fact that all Monarch butterflies share the exact same color pattern on their wings suggests that this specific coloration and design must be "written" into this species' genetic blueprint–into its genes. How else are we to explain the uniformity of a Monarch's wings? Are we to believe that they all possess this same intricate pattern as the result of some vast coincidence, as if Monarch butterflies can be born with any combination of colors on their wings and in any design, only it just so happens that, by pure chance, they always turn out the same? Hardly! No less of a coincidence than the fact that all fish have gills or all cats have whiskers, all Monarchs possess the same elaborate design and color pattern on their wings. Apparently, the Monarch's unique display exists as the result of information encoded in that species' genes.

This can be said of all universal characteristics possessed by any species. Whether we are discussing a butterfly's wings, a rat's tail, or a human's brain, each represents a physical characteristic that emerges as the result of information stored within that species' genes. We could therefore say that, as a rule, for every physical characteristic that is common to every member of a given species, there must exist genes responsible for the emergence of that trait.

Not only does this rule apply to universal physical features but to universal functions as well. For instance, all humans grow hair. The fact that hair growth represents a universal characteristic of our species suggests that this must represent a genetically inherited trait. Because growing hair is a function, it implies our species must possess some specific set of genes that instruct our developing bodies to forge specific physiological sites in us, some mechanism from which hair growth will be generated. In this particular case, such sites are represented by our hair follicles. Unless we are to believe that hair magically appears from our skin, it is necessary that there exists some physiological mechanism responsible for hair production.

This suggests that for every universal function, be it an organism's capacity to smell, hear, see, breathe, ingest, digest, reproduce, etc., two things must be true: one, for every inherited function, there must exist some specific physiological site or set of sites from which that function is generated; two, there must exist some underlying gene or set of genes responsible for the emergence of those physiological sites that perform that function.

According to the science of sociobiology, the above principle can also be applied to universal behaviors. Take, for instance, the movements of the planarian, a creature belonging to the phylum of flatworms (platyhelminthes). Planarians do not have brains but instead have several longitudinal nerve cords that run the length of their tiny bodies to a head where these few nerves converge. Rather than refer to this cluster of nerves as a brain, it is called a cephalic ganglion, constituting a relatively simple central nervous system.

Every planarian has a distinct tendency to maneuver its body in such a way that its head is always turned in the direction of a light source, a phenomenon referred to as phototaxy. That all planarians engage in this specific phototactic behavior implies that it represents a universal characteristic of the species.

Analogous to a Monarch's design, there are three possible reasons all planarians respond to light in this particular way. The first is that all planarians turn toward light because they are taught to do so by others of their species. In other words, perhaps phototaxy is a learned behavior. The problem with this explanation is that should we isolate any single planarian from the moment of its conception from all others, allow it to develop to adulthood, and then place it in a space with a light source at one end, it will invariably turn in that direction, implying that phototactic behavior is not one that needs to be learned by this species.

The second possible reason all planarians orient themselves in the direction of light is that they want to do so–they do it as an act of free will. As we can never truly know what a planarian is "thinking," we can never know whether this is the case or not. Nevertheless, if planarians did have the wherewithal to make such free and voluntary decisions, what are the chances that every single one of them would always choose to behave in this same exact way, moreover, all the time? Wouldn't it be reasonable to presume that some might choose to turn away from the light, even if just some of the time? Are we to believe that all planarians always engage in this same propensity as the result of some vast coincidence, as if each individual within the species actually possesses free will, and that any day now they might all suddenly change their minds and decide to turn away from the light? Again, highly unlikely. That all planarians always turn towards light leads me to believe that this is not a case of either free will or coincidence.

The third possible reason that all planarians exhibit this phototactic response is that infused within the planarians' ganglion, there exist genetically inherited neural connections that compel every member of the species to respond to light in this particular way, thus implying that phototactic behavior represents a genetically inherited reflex.

So, which of these various possibilities are we to believe? As incredulous as the first two might seem, one cannot base a theory strictly on the process of elimination. If we are to speculate that planarians orient themselves towards the light because they are genetically hardwired to do so, we need positive confirmation.

Planarians perform this phototactic feat by continuously shifting their bodies until the two light receptors (what we would call their eyes) situated in their cephalic region (their head) are equally stimulated. In experiments performed on the species, it was found that "if two equally bright lights a short distance apart are placed near the planarian, the animal will orient itself toward a point midway, thus attaining equal stimulation of the two eyes."[4] The fact that a planarian's movements can be manipulated in such a way and with such consistency would attest to the fact that planarian phototaxy represents the consequence of a physiologically generated reflex–not free will and not coincidence.

As more evidence to support a neurobiological explanation of planarian phototaxy, in an article published in the *Journal of Experimental Biology*, "Rhodopsin-like proteins in planarian eye and auricle: Detection and Functional Analysis," it was discovered that when a specific rhodopsin-like protein located in the planarian's cephalic ganglion was removed, the animal was no longer responsive to light. Most revealing, not until days later, after these proteins had regenerated, was the creature's phototactic reflex restored. Based on this observation, it was concluded that "rhodopsin-like proteins in the eyes work as photoreceptors for phototactic behavior."[5]

The fact that planarian behavior can be reduced to chemical processes confirms that this organism responds to light neither as an act of free will nor as a learned behavior, but rather as the result of a completely involuntary physiological response to a specific stimulus, again, a reflex. Similar to the way we can electrically hardwire a mechanical device to turn towards light, nature* has hardwired planarians with this same propensity. What this suggests is that universal behaviors represent genetically inherited traits. No differently than Monarch butterflies inherit their unique wing design, planarians inherit their phototactic reflex.

What if we were to apply the same principle to a more advanced species? Consider, as another example, the fact that all honeybee colonies construct their honeycombs in the same hexagonal fashion, regardless of whether or not they've ever been exposed to another honeybee colony. When bee larvae are removed from their colonies and raised under artificial conditions, they still emerge as adults to construct their hives with the same hexagonal design. If we were to apply the same principle to the honeybees that we did to planarians, it would

---

*I would like to make it clear that when I refer to "nature" as a force of evolutionary change, I do not mean to imbue it with any sense of consciousness, will, intelligence, awareness, or intent. In essence, I'm really just using the word as a metaphor for the laws of thermodynamics–those underlying physical principles to which all matter and energy are inexorably bound and which have therefore determined all that has transpired in the physical universe from the moment of its conception. Though proponents of intelligent design believe that the unfolding universe is simply too complex to be a mere series of fateful physical accidents and that there must therefore exist some conscious entity that oversees and intervenes with all that occurs, I do not share their assessment as it is purely faith-based and has no grounding in science whatsoever.

imply that bees construct their hives in adjoining hexagons as the result of a genetically inherited reflex.

Moving along the phylogenetic ladder, how about the fact that all three-spined sticklebacks, a species of fish, perform the same species-specific zig-zag dance as part of their courtship and reproduction ritual. To confirm the innate nature of this behavior, the ethologist R. A. Hinde performed a series of deprivation experiments in which newly hatched sticklebacks were reared in complete isolation without any exposure to other members of their species. Before the male stickleback builds a nest for his future mate, he clears the area of all potential competition by chasing away all other male sticklebacks (identifiable by their red bellies). What Hinde found was that "male sticklebacks raised in isolation will attack a red-bellied wooden model even though they have never seen a male stickleback before."[6]

As another example of a built-in reflex, we can point to the case of the herring gull. When a newly hatched herring gull chick pecks on its mother's beak, the mother will instinctively regurgitate food from its crop to feed her young. In order to study the inherent nature of this behavior, the Nobel Prize–winning ethologist Niko Tinbergen offered newly hatched gull chicks various cardboard models of gull heads and observed which of them elicited the greatest response. What Tinbergen found was that of all the cardboard models he placed before the chicks, the one they pecked at the most was the one with the long, thin red beak characteristic of an adult herring gull. What made this so particularly revealing was the fact that the chicks had never had any exposure to an adult gull before, thus confirming that the chicks were genetically hardwired with information that enables them to recognize an adult of its species (not unlike a human infant's capacity to innately recognize and then suckle on its mother's nipple). This would suggest that somewhere in the herring gull chicks' brains there exists a series of neural connections that compel them to engage in this behavior. We could therefore say that herring gulls possess a "pecking" part of their brain. Sever these connections and it's unlikely that herring gull will any longer be able to enact this reflex.

Moving further along the phylogenetic trail, how about the fact that all cats meow? Take a kitten away from its mother, for instance, and raise it in total isolation, and it will still meow, suggesting that meowing is an inherited reflex. This would further imply that there must exist a "meow" part of a cat's brain from where this capacity is generated. Disable this cluster of neural connections and, in all likelihood, that cat will lose its capacity to meow. Furthermore, this would equally imply that cats must possess what we could call "meow" genes which are responsible for the emergence of those neural connections that make up the meow part of the cat's brain.

Moving along to primates, how about the fact that all Eastern Mountain gorillas engage in the same species-specific play behaviors, courting and reproductive rites, foraging and child-rearing techniques, and threat and submission displays, to name just a few of their universal propensities? How is it possible that all troops belonging to this species–regardless of whether they've been exposed to one another–engage in such similar behaviors? Are we to believe that the species is psychic and telepathically communicates its behaviors to other troops across the plains? Or is it because since all Eastern Mountain gorillas exist as a part of the same species and therefore possess the same genes, they are all hardwired to behave in similar ways? Just as all planarians turn towards the light, all gorillas engage in species-specific play, grooming, foraging, and courtship behaviors. Does this mean that primate behavior, similar to that of a planarian, bee, herring gull, or cat, can be summarized as the consequence of a series of inherited reflexes? As the biologist William Keeton asked:

> Should we then regard reflexes as the fundamental behavioral units? In a sense, yes...And it is true that there is no difference between simple reflexes and more complex reactions; every conceivable intermediate stage exists between the simplest reflex pathway and the most complicated neural pathway.

> It is possible to view even the most complex behavior as the result of an intricate interaction among many enormously complex reflexes.[7]

Suppose we were to climb even higher along the phylogenetic ladder, all the way to Homo sapiens. Shouldn't these same biological principles that apply to every other life form apply to the human animal as well? Well, science does apply these same principles to humans and has noted quite a few universal behavioral patterns (in the case of humans, what are referred to as cross-cultural behavioral patterns) in our species as well–behaviors that have been exhibited in some form by every culture from the beginning of our species. As the social critic Ralph Linton eloquently expressed this notion:

> The essential unanimity with which the universal cultural pattern is accepted suggests that it is not a mere artifact of classificatory ingenuity but rests upon some substantial foundation. This basis cannot be sought in history, in geography, or race, or any other factor since the universal pattern links all known cultures, simple and complex, ancient and modern. It can only be sought, therefore, in the fundamental biological nature of man and in the universal conditions of human existence.[8]

If all planarians orient themselves towards light, they must be genetically preprogrammed to do so. If all Eastern Mountain gorillas engage in species-specific courtship rites, they, too, must be genetically programmed to behave this way. Whether we like to believe it or not, humans are animals, too. Therefore, whatever logic applies to all of the Earth's other creatures must also apply to our own. If there is any behavior that has been universally exhibited among every human culture, it suggests that in all likelihood that behavior constitutes a genetically inherited instinct as well.

Take, for instance, the fact that humans from every culture express the emotions of grief, fear, aggression, and amusement with the same exact facial expressions.* For example, all humans express the sentiment of amusement with a facial expression we refer to as a smile. Even the blind, who have never seen another person, smile when amused, thus confirming the reflexive nature of this fundamental human expression. That all humans express amusement in the same exact way suggests that, just as all planarians turn to light, all humans express their emotional states as the result of completely involuntary, genetically inherited reflexes.

With this in mind, let us investigate some other, more complex cross-cultural behavioral patterns evident in the human animal, behaviors that have been found to exist among every human society from the dawn of our species. Some examples of such cross-cultural patterns include the arrangement of kin-groups; the application of sexual restrictions; birth, puberty, marriage, and death rites; acts of celebration, mourning, and courtship; incest taboos; inheritance rules; weaning; education of young; hygiene; obstetrics; status differentiation; division and cooperation of labor; community organization; development of legal codes and penal sanctions; toolmaking; trade; cooking; gift-giving; joking; use of personal names; playing games and sports; dancing; singing; religious worship; creation of musical instruments; bodily adornment; use of calendars; counting; belief in magic and the supernatural; medicine; mythology; government; and language.

Does this mean that our species is genetically predisposed to engage in such seemingly abstract behaviors as the application of math, language, music, or even religion? Is it possible that such behaviors could exist as the consequence of a genetically inherited impulse or instinct? Let's look at language, for example. Among cultural anthropologists and linguists alike, it's agreed that all human cultures communicate through a spoken language. Because

*As an exception to this rule, people born with a damaged or dysfunctional fusiform gyrus, that part of the brain from which we derive our capacity to distinguish certain facial cues, will not possess this expressive capacity.

we all possess this linguistic capacity, we can assume that it represents a genetically inherited characteristic of our species. This would further imply that there must exist physiological sites in us from which these language capacities we possess are generated. Moreover, this would also suggest that we must possess what we could call "language" genes responsible for the emergence of any such language parts of the brain.

So where does linguistic intelligence originate? Does it stem from our hearts, our kidneys, our livers? Of course not. Like all cognitive capacities, ours for language originates from within the brain. How do we know this? We know this because there is physical evidence to prove it.

Within the human brain (and only the human brain), there exist specific structures responsible for the generation of our language capacities. Such language-enabling parts of the brain include the Broca's area, Wernicke's area, and the angular gyrus. The angular gyrus is the part of our brain that receives sensory information such as the scent of a flower, the taste of a lemon, or the sound of a bell, and then links that sensory input to its verbal correlate or "word." For example, when we smell a rose, our angular gyrus recalls the word "rose" prompted by the scent. The angular gyrus therefore acts as our brain's linguistic filing cabinet, that place where all the words through which we've learned to define our sense-experiences are stored.

The Wernicke's area, which is located in the brain's temporal lobe and plays an essential role in linguistic comprehension, retrieves the recalled word from the angular gyrus and then processes it in such a way that we can grasp that word's meaning. From there, the Broca's area, which controls the muscles of the face, jaw, palate, and larynx, allows for our words to be physically spoken.

And how do we know these organs exist in us? We know this because in cases where any one of these sites has incurred physical damage, it has been shown to have a direct effect on some specific part of that person's language abilities. Such linguistic malfunctions are known as aphasias. Damage, for example, incurred to one's Wernicke's area, which is vital to comprehension, can effect a person's ability to comprehend the meaning of words they previously understood. In some cases, damage can be so specific that though a person might not be able to comprehend a word when it is heard, he will comprehend the meaning of that same word when it is written. In other instances, damage to Wernicke's area can produce speech that, though it may be fluent, will be meaningless.

Damage to Broca's area, which controls articulation, will cause impairment of speech so that articulation may be slowed, labored, or completely disabled, depending on the extent of the injury. In some cases, the damage can be so specific that while a person may be able to say the word "hopper," for example, he will not be able to say the word "hop." As we can see, which specific part of our language center is damaged determines what specific language dysfunction a person will suffer.

Similar to the manner in which removing a specific part of a planarian's ganglion will affect its phototactic response, if we damage or remove a specific part of one's language center in the brain, it will affect that person's linguistic response. Just as a planarian's behavior can be reduced to electrochemical processes, the same is apparently true for our species as well.

What all this demonstrates is that there exist very specific physiological sites within our brain that are responsible for our specific language and speech capacities. No less than we all possess two eyes, ten toes and a heart, we all possess an angular gyrus. And again, how do such physiological sites emerge in us? From information stored within our genes. Just as we possess genes that instruct our emerging bodies to develop a heart within our thoracic cavity, we possess genes that instruct our emerging bodies to develop an angular gyrus within our brain.

Moreover, just as our capacity to speak and comprehend a language was passed on to us through our parents' genes, we will pass this same capacity on to our own offspring. In other words, cognitive traits are no different from all other physical traits in that they are passed from generation to generation through the transmission of genetic material. Just as such basic physical attributes as eye or skin color are predetermined by genetic inheritance, the same is true for our inherent language capacities. Furthermore, just as our language capacities are genetically conceived, the same is likely to be true of all of our cross-cultural propensities.

How about music as yet another example of a cross-cultural behavior in our species? No plant, insect, fish, cat, dog, or even chimp uses either its body parts or various materials to create rhythmic combinations of sound. Humans, however, do. As a matter of fact, every human culture that has ever existed has demonstrated a capacity for music. Does this mean that something as inspirational as musical creation might exist as the effect of a genetically inherited reflex? Is it possible that Mozart's talent may have represented the physical consequence of his being born with enhanced "musical" genes? Perhaps, for if music does indeed represent a cross-cultural characteristic of our species, it would suggest that there must exist a "musical" part of the brain from which this capacity is generated. And what evidence might there be to support such a notion? According to the musicologist John Blacking:

> There is so much music in the world that it is reasonable to suppose that music, like language and possibly religion, is a species-specific trait of man. Essential physiological and cognitive processes that generate musical composition and performance may even be genetically inherited and therefore present in almost every human being.[9]

It is generally agreed that every human culture from the beginning of our species has generated some form of music. "No culture so far discovered lacks music."[10] This would imply that if I clapped my hands in a rhythmical manner in the company of almost anyone from any culture, there exists a distinct possibility that he or she would have the inclination as well as the ability to join in with me. As we know, this is not something I could achieve with a plant, insect, fish, cat, or any other animal. Expressing oneself musically is, therefore, an exclusively human capacity.

In addition to the fact that music has emerged from every culture, what other evidence is there to support this notion that we might possess a "musical" part of our brain? Let's take the capacity known as perfect pitch. Here is an aptitude some people possess with which they can determine the exact pitch of any sound they hear. But perfect pitch cannot be learned. One must be born with it. This implies that the capacity for perfect pitch is innate.

What about musical "idiot savants," people born with incredible musical abilities who are intellectually retarded in almost every other way; people, for example, who after hearing a complete Beethoven sonata for the first time, can sit down at a piano and play the same piece, note for note and in perfect time, but meanwhile can't tie their own shoelaces? We hold musical talent in such high esteem, as one of the trademarks of human genius and inspiration. In light of the "idiot savant," however, is this an act of an inspired genius or something more mechanical in nature, perhaps the consequence of a genetically inherited instinct–a sophisticated reflex? If we can create machines that can play music, why should it be so difficult for us to believe that nature could have designed us to do the same?

How about the fact that people can suffer from musical aphasias? Similar to a linguistic aphasia, musical aphasias constitute the loss of some specific musical ability caused by physical damage incurred to one's brain. For example, after suffering a stroke, a composer may lose his ability to write music; a musician, his ability to play an instrument. That musical aphasias exist suggests that, just as with language, our musical abilities must be integrally linked to our neurophysiological makeups.

How about the fact that music can affect us physiologically? "Music can provoke intense, genuine, emotional arousal from ecstatic happiness to a flood of tears."[11]

Equally revealing is the fact that, regardless of one's cultural origins, all peoples tend to interpret certain musical themes in the same way. Who, for instance, and from what culture, would ever describe a John Philip Sousa march as soothing or tranquil, as opposed to militant, triumphant, or exhilarating? Wouldn't the fact that people from different cultures experience and interpret the same musical stimuli in similar ways suggest that musical consciousness must represent an inherent part of our species' physiology?

Another phenomenon which may show that our musical capacities are physiologically based is the fact that certain combinations of sounds have been shown to trigger epileptic seizures. "Musical epilepsy convincingly demonstrates that music has a direct effect upon the brain."[12]

Without getting any more deeply involved in an argument supporting the existence of a "musical" part of the brain, it seems there exists adequate evidence to suggest that our capacity for music is directly linked to our cerebral physiologies. What this means is that music and language represent two ways that the human brain inherently processes information, two of the many ways that our physiological makeups determine the manner in which our species interprets reality.

After having acquired what I felt constituted adequate evidence that cross-cultural behaviors represent the effects of genetically inherited impulses, it was now time to apply this same principle to humankind's cross-cultural propensity to believe in a spiritual reality. Just as every culture from the dawn of our species has perceived the world musically and linguistically, every culture has perceived the world spiritually.

Was it therefore possible that humans may actually inherit their cross-cultural inclinations to perceive a spiritual reality? Were our cross-cultural beliefs in such universal concepts as a god, a soul, and an afterlife the consequence of a genetically inherited instinct, a reflex? Furthermore, if we possess such an instinct, mustn't it emerge from some specific physiological site in us, what we could perhaps call a "spiritual" or a "God" part of our brain?

Book II

Intro to Biotheology

"The heart has its reasons, which reason does not know. We feel it in a thousand things. I say that the heart naturally loves the universal being."

–Pascal

"It seems that the existence of God is self-evident. Those things are said to be self-evident to us the knowledge of which is naturally implanted in us."

–Thomas Aquinas

"The predisposition to religious belief is the most complex and powerful force in the human mind and in all probability an ineradicable part of human nature."

–E. O. Wilson

CHAPTER 7

THE "SPIRITUAL" FUNCTION

"All the civilizations of mankind that have existed were rooted in religion and a quest for God."[13]

–IVAR LISSNER

Every generation of every* human culture, no matter how isolated, has possessed the capacity to speak and comprehend a language. This suggests that within our chromosomes there must exist genes from which our linguistic capacities emerge in us. As we develop within the womb, it is the role of these "language" genes to

*I would like to qualify the use of the word *every* when I make such sweeping statements as to refer to "every world culture." More precisely, what I'm referring to is every world culture that has been properly observed and recorded by the world's preeminent cultural anthropologists. Nevertheless, it needs to be stated that there have existed a myriad of cultures, now extinct, who were never witnessed by outsiders or, if they were, who were never properly documented and therefore cannot be accounted for. It is also not to suggest that though the vast majority of human societies have most likely conformed to these assumptions, it is within the realm of possibility that there may have existed cultural anomalies throughout our history who have defied such seeming rules of human nature.

instruct our developing bodies to forge specialized neurophysiological connections that will one day constitute those sites from which our linguistic capacities will be generated. As a matter of fact, for every behavior that is universal to any species, there must exist specialized genes that prompt the development of specialized neurophysiological sites from which those behaviors are generated.

What if we were to apply this same principle to human spirituality? Just as all human cultures have demonstrated a propensity to develop a language, all human cultures have just as clearly demonstrated a propensity to develop a religion as well as a belief in a spiritual reality. According to the Pulitzer Prize–winner E. O. Wilson:

> Religious belief is one of the universals of human behavior, taking recognizable form in every society from hunter gatherer bands to socialist republics. Its rudiments go back, at least, to the bone altars and funerary rites of Neanderthal man.[14]

As asserted by such men as Carl Jung, Joseph Campbell, and Mircea Eliade, every world culture from the dawn of our species has maintained a dualistic interpretation of reality–every culture has perceived reality as consisting of two distinct substances or realms: the physical and the spiritual. Accordingly, objects that belong to the physical realm are viewed as tangible, corporeal, that which can be empirically experienced or validated (i.e., seen, felt, tasted, smelled, or heard). Objects that exist as a part of this realm are subject to the physical forces of change (i.e., birth, death, and decay), and are consequently perceived as existing in a state of constant flux, temporal, fleeting.

On the other hand, our species equally perceives the existence of a spiritual realm. As this realm transcends the nature of the physical/material universe, things comprised of spirit are immune to the laws of physical nature (i.e., to change, death, and decay). That which exists as a part of the spiritual realm is consequently perceived as being permanent, fixed, eternal, everlasting.

As confirmation of the cross-cultural nature of man's dualistic interpretation of reality, every culture from the dawn of our species has maintained a belief in the existence of unseen spirit guardians we refer to as gods. According to Dr. Herbert Benson, "There is not a civilization known to us that did not have faith in God or Gods."[15] Man, the musical animal, the mathematical animal, the linguistic animal, is also the spiritual animal. Now if it's true that all cross-cultural behaviors represent genetically inherited traits, then shouldn't we presume that the same must hold true for our species' propensity to believe in a spiritual reality? Wouldn't the fact that all human cultures, no matter how isolated, have believed in the existence of a spiritual realm suggest that such a perception must constitute an inherent characteristic of our species, a reflex?

If we inherit our spiritual proclivities and beliefs, wouldn't this further imply that we must possess genes through which this instinct to believe is passed from one generation to the next? Furthermore, if we inherit our propensity to believe in a spiritual reality, mustn't there exist some physiological site in us from which these "spiritual" perceptions, sensations, and cognitions are generated? Since all perception, sensation, and cognition originates from within the brain, it follows that "spiritual" consciousness must be generated from that same organ. Consequently, if believing in a spiritual reality represents a cross-cultural characteristic of our species, this would imply that we possess a neurophysiologically based "spiritual" function or what I informally refer to as the "God" part of the brain.

Jung

As I began to explore the possibility that we may inherit our spiritual proclivities, I found there were others who had already made similar inquiries, others from whose work and research I could now borrow. Of those who had conducted such studies, it was the work of the analytic psychologist Carl Jung I found most pertinent. Of all Jung's contributions, however, it was his theory of the "collective unconscious" I found most applicable.

Jung's mentor, Sigmund Freud, had introduced the concept of a personal conscious and unconscious to the world. According to Freud, the personal conscious represented those thoughts, feelings, memories, and desires of which we are consciously aware. Beneath the personal conscious lay an even deeper layer of consciousness represented by an individual's unconscious self. According to Freud, one's primal drives or instincts, their personality components, memories of early childhood experiences, repressed memories, and other inner conflicts all reside within one's personal unconscious. Though we might not be aware that these elements exist in us, they nonetheless play a significant role in all that we do, say, and think. To Freud, the personal conscious and unconscious represented the two chief components underlying all human behavior.

Jung picked up where Freud left off (something for which Freud supposedly never forgave him) by suggesting there existed an even deeper and more profound layer of human consciousness than that of the personal unconscious. Jung maintained that beneath the personal unconscious and acting as its foundation there existed what he referred to as the collective unconscious.

According to Jung, whereas the personal conscious and personal unconscious are derived from one's personal experiences, the collective unconscious represents those components, awarenesses, and drives that we inherit and which therefore constitute an integral part of the conscious experience that is mutually shared by every member of our species. Whereas the contents of one's personal unconscious emerge from one's personal experience and development, the contents of one's collective unconscious constitute that part of us which was forged during our species' development and which is therefore common to all humankind. The collective unconscious therefore exists as a part of our inherent natures and "has contents that are, more or less, the same everywhere and in all individuals. It is, in other words, identical in all men and constitutes a common psychic substrate of a suprapersonal nature which is present in every one of us…and which has existed since the remotest times."[16]

Whereas the philosopher John Locke believed we are born as a "tabula rasa," a clean slate, waiting for our experiences alone to shape and define us, Jung, in accordance with Kant, held that we are born with a set of preprogrammed modes of perception. Like Kant, Jung seemed to be directing his search for answers inward, into the nature of human consciousness.

Jung reached many of his conclusions based on comparative studies he made of the world's various mythologies. These mythologies, he found, each constituted a similar compilation of fables, legends, and morality tales that exist among every human culture from the dawn of our species. Through its mythology, every human culture has codified its social and spiritual norms, rites, customs, ethical standards, and beliefs. Jung not only concluded that all cultures possessed a mythology, but that all of them also contained remarkable similarities. Whether he was studying the Old and New Testaments of Judeo-Christianity, the Zarathustrian Avestas, the Norse Eddas, the Icelandic Sagas, the Islamic Koran, the Egyptian or Tibetan Books of the Dead, Hesiod's *Theogony*, Homer's *Iliad* and *Odyssey*, Virgil's *Aeneid*, the Celtic Sagas, Urartian (Armenian) cuneiform, the Japanese Kojiki (Record of Ancient Masters) or Nihongi (the Chronicles), the Babylonian tales, the Ugaritic myths of Palestine and Syria, the Chinese Shi Ching (Book of History), the Hindu Rig Veda, Mahabharata and Ramayana, the Theravada Buddhist Vinanatthu, the myths from the various cultures of Africa, Polynesia, or South and Central America, or the manuscripts of the medieval Alchemists, Jung found common themes in each of these culture's writings.

Because he found such similarities in the myths of every world culture, Jung concluded that the contents of these myths must be generated from some inherent psychic substrate that must be shared by our entire species. This he called our collective unconscious. Apparently, our species possessed some impulse that not only prompted each culture to create its own mythology but that fashioned each with the same universal themes. Jung referred to such common themes as archetypes. Due to the universal nature of these archetypes, Jung postulated that our species possessed an inherent religious function:

> Through the study of the archetypes of the collective unconscious we find that man possesses a religious function and that this influences him in a way as powerfully as do the instincts of sexuality and aggression. Primitive man is as occupied with the expression of this function, the forming of symbols, and the building up of religion as he is with tilling the Earth, hunting, fishing, and the fulfillment of his other basic needs.[17]

Inspired by Jung's theories, particularly by his suggestion that humans possess what he called a "natural religious function," I now felt, more than ever, that human beings might indeed inherit their spiritual sensibilities. The chief difference between my interpretation and Jung's, however, was that whereas Jung perceived human consciousness as a function of the human mind, I saw it as a function of the brain. Whereas the mind implies the possibility that there might exist some intangible, transcendental component within us, the brain does not. Whereas advocates of the "mind" perceive cognition to be a function of a soul, a ghost in the machine, advocates of the brain perceive cognition to be a function of one's neurophysiology, a mechanical reflex.

Based on what the neurophysiological sciences had taught me (sciences that simply weren't available to Jung in his time), I had adopted a more rational, mechanistic–scientific–approach to human sensation, perception, emotion, and cognition, that is, to the contents of human consciousness.

What if I were to apply these newly advanced neurophysiological sciences not just to consciousness but, more specifically, to Jung's notion of the collective unconscious? What if it were possible to "biologize" the collective unconscious, to reduce it to neurochemical processes? What if what Jung spoke of as a natural religious function could be explained as a genetically inherited predisposition? By applying the neurophysiological sciences to the study of human spirituality, I now felt it might be possible to construct a purely mechanistic–a scientific–interpretation of human spirituality as well as of God.

Universal Spiritual Beliefs and Practices

"The history of religion–from the most primitive to
the most highly developed–is constituted by a
great number of sacred realities."[18]
–MIRCEA ELIADE

As I explored the world's various cultures–each with its unique beliefs and practices–it became apparent that each had maintained a dualistic interpretation of reality, each had perceived reality as being comprised of two distinct realms: the physical and the spiritual. If, indeed, there was substantial evidence to support the argument that spiritual belief was truly a universal, then, according to the principles of sociobiology, it followed that it was highly likely that we, as a species, must be "hardwired" this way. So how prevalent is religious and spiritual belief? Is there ample evidence to suggest it's all part of an inherited reflex?

The universality with which we perceive a spiritual reality is manifest in a number of cross-cultural beliefs and practices. For instance, every culture has expressed a belief in supernatural forces or beings. This is made evident by the fact that every culture has demonstrated a tendency to pray to, worship, and petition such beings, most commonly referred to as gods–a concept for which every culture has possessed a symbol or word. This is further supported by the fact that every culture has erected sites of worship through which the members of its community can gather to pray to its gods. Whether it be a Muslim mosque, a Catholic church, a Jewish synagogue, a Shinto shrine, a Babylonian ziggurat, a Buddhist stupa, or an ancient Aztec, Greek, or Egyptian temple, every culture has constructed physical edifices specifically designed for the sole purpose of praying to and petitioning one's gods. Such sites of worship constitute physical evidence that all cultures have believed in the existence of a spiritual reality.

In addition, every culture has created religious works of art. The first examples of this exist in the form of cave paintings which date as far back as to man's early Paleolithic age, from about 40,000–12,000 BCE. These early cave paintings often depict representations of a hunt in which various animals are covered with javelin wounds highlighted with red ochre. Because the spear designs were often painted over one another, it is believed that these paintings were constantly renewed for magico-religious purposes to help effect a kill in the chase. In written form, every culture has expounded upon its spiritual beliefs through scriptures and mythologies. As a matter of fact, the Sumerians, who devised one of the first systems of written communication (circa 2,800 BCE) in the form of inscriptions known as cuneiform, had, among some of their first symbols, a sign ("an") that stood for heaven. That all cultures have possessed such tangible works of art and text constitutes further evidence that the human animal cross-culturally perceives and believes in a spiritual reality.

Furthermore, every world culture has maintained a belief that humans possess a spiritual component that exists within us, what is otherwise referred to as a soul–another concept for which every culture has possessed either a symbol or word. "The soul is a universal concept."[19] According to our cross-cultural belief in a soul, humans perceive themselves as being comprised of a unique combination of both matter and spirit. While we perceive our bodies to be constituted in matter, we, at the same time, perceive consciousness as being constituted in spirit, an intangible substance we refer to as one's soul. In this way, we project our dualistic perception of reality onto our own existences.

Just as we perceive things that consist of spirit as being indestructible, eternal, and everlasting, we perceive our souls as possessing these same attributes. Consequently, we believe that by virtue of our souls, we, our conscious selves, are eternal and everlasting. As a result, we believe that though our physical bodies will one day perish, our spiritual self–our spirit or soul–will persevere for all eternity. It is through this universal belief in a soul that human beings derive their sense of immortality. In the words of the sociologist Branislaw Malinowski:

> Through religion man affirms his convictions that death is not real nor yet final and that we are endowed with a personality which persists even after death.[20]

The universality by which all cultures have believed in an immortal soul is supported by the fact that all cultures have expressed a belief in an afterlife, "a new, continued or transformed existence after death, belief in which has been found in virtually all cultures and civilizations."[21] Be it Heaven, Purgatory, Hell, Valhalla, Niflheim, Nirvana, Tartarus, the Elysian Fields, Hades, Oblivion, the Realm of the Dead, the spirit land (Te Reinga), the Mystical Garden, Paradise, reincarnation, or transmigration of the soul, all cultures–Eastern and Western–have expressed a belief that our spiritual selves or souls persist long after our physical bodies have perished.

This universal belief in an afterlife is physically manifest in the cross-culturally enacted funerary or burial ritual. In this universal practice, the deceased's body is disposed of (generally buried, though there are other means) with a rite that anticipates sending that individual's spirit on to some next or other realm. As further physical evidence, many cultures bury their dead with artifacts meant to facilitate the deceased's transition from this realm to the next, providing yet more confirmation in these culture's beliefs that our conscious self or soul endures after physical death.

While burial represents the last of a series of cross-culturally enacted rituals through which we sanctify our existences before our gods, all cultures inaugurate the newly born into their spiritual community with a birth rite. Examples of such rites include a Jewish or Muslim circumcision, the immersion of a Catholic child into the baptismal font, or the Australian Aborigine rocking its newborn through the purifying smoke of the Konkerberry fire. As the cultural anthropologist Mircea Eliade expressed in his landmark work, *The Sacred and the Profane*, "When a child is born, he has only a physical existence; he is not yet recognized by his family nor accepted by the community. It is only by virtue of those rites performed immediately after birth that he is incorporated into the community of the living."[22]

The next life passage, after birth, that is cross-culturally addressed in a spiritual context comes in the form of an initiation rite. This rite, which is usually celebrated in tandem with puberty, signifies one's passage from childhood to adulthood and is meant to sanctify an individual before his gods as a grown and responsible member of the spiritual community. Whether it be a Jewish Bar Mitzvah, a Congolese Kota face-painting ceremony, a Catholic Confirmation, an adolescent baptism of the Southern Baptist, or a Hindu Sannya ceremony, every culture performs a ritual by which it assimilates its young members into the spiritual community as an adult. Using Jungian terms to express the cross-cultural nature of this rite, the author Anthony Stevens writes in his book *On Jung*, "Comparison of rites from all over the world suggest that these initiation rites themselves possess an archetypal structure, for the same underlying patterns and procedures are universally apparent."[23]

After being initiated into the spiritual community, members of the opposite sex are united to promote procreation. Such unions are spiritually sanctioned through a cross-culturally practiced marriage rite.

In addition, every culture has possessed some form of a priesthood, some individual or group of individuals whose role is to act as the community's intermediary between the material and spiritual worlds. Whether this individual is referred to as a shaman, priest, rabbi, swami, ensi, yogi, oracle, mystic, psychic, medium, pope, caliph, or imam, all cultures have possessed some such member, group, or caste whose role is to serve as its community's spiritual guide and leader.

Moreover, all cultures ascribe magical, sacred, or supernatural–spiritual–status to certain locations, what Mircea Eliade refers to as our species' tendency to believe in the notion of "sacred" space. For example, every culture has ascribed sacred status to a number of sites referred to as shrines. Whether it be the Tomb of the

Patriarchs, the Kaaba Stone, Delphi, the Pyramids, the Dakhma of Cain, the Ganges River, Bethlehem, or a Buddhist Stupa, each represent centers of pilgrimages and adoration because of their spiritual significance and the spiritual values they've come to symbolize.

Sacred status has also been cross-culturally ascribed to various objects. Totems, relics, icons, amulets, talismans, charms, or fetishes, as they are called by their respective cultures, all represent examples of physical objects believed to contain some essence of the spiritual realm within them. Whether it be the wafer and wine of the Eucharist, the ceremonial Calumet or Peace Pipe of the Native Americans, the hairs of the prophet Mohammed, the sacred tooth of Buddha, fragments of the holy crucifix, a mezuzah, an African gris-gris, or an amethyst or quartz crystal for the "new age" spiritualist, all represent material objects that are believed to possess magical or "spiritual" attributes. That all cultures have assigned such sacred status to physical objects further attests to the fact that all human cultures have maintained a belief in the existence of a spiritual reality.

Furthermore, all cultures have expressed a belief in the existence of spiritual/transcendental/supernatural forces that guide and influence all that transpires in our world. This is made evident by our beliefs in such abstractions as luck, karma, kismet, fate, fortune, and destiny. Such concepts demonstrate our perception that there exist transcendental forces which influence and intervene in all that occurs within the material universe. In the same vein, all cultures exhibit superstitious behaviors in which they believe that certain gestures (e.g., crossing one's fingers, knocking on wood, throwing salt over one's shoulder) or charms (e.g., a holy cross or rabbit's foot) can help bring us luck which, in essence, is the belief that we can alter the course of destiny by appealing to some supernatural force or realm.

Another cross-culturally enacted behavior that attests to man's inherent propensity to believe in a spiritual reality is necromancy, the belief that we can communicate with the spirits of the dead.

This coincides with our species' tendency to believe in ghosts, the phantasmagoric incarnations of those who have passed away.

Humankind's universal belief in a spiritual element is further evinced by the fact that all cultures tend to associate the sentiment of guilt in a religious context. Though we may feel guilty for things we've done to other men, all cultures show an express concern for how their actions will be judged by their gods. This is made evident by a variety of rites of atonement and penitence through which individuals from every culture have sought to repent for crimes committed against their gods. Such crimes are known as sins, another concept for which every culture has possessed a word.

Physical evidence of penitent behavior is manifest by a variety of sacrificial rites. In these rites, individuals make offerings to their gods in the hope that it will solicit their sympathy, mercy, or forgiveness. We engage in acts of penitence because we believe such acts will be rewarded by our gods both in this lifetime as well as in the afterlife.

To provide a concise example of how some of the above sentiments have been expressed through one culture's sacred literature, I will turn to the Sumerian text *The Counsels of Wisdom* (135-145):

> Worship your god every day, with sacrifice and prayer which properly go with incense offerings. Present your freewill offering to your god for this is fitting for the gods. Offer him daily prayer, supplication, and prostration and you will get your reward. Then you will have full communion with your god. Reverence begets favor. Sacrifice prolongs life, and prayer atones for guilt.

The Argument For a Spiritual Function

"If humankind evolved by Darwinian natural selection, genetic chance and environmental necessity, not God, made the species."[24]
–E. O. WILSON

All human cultures have practiced a belief in the existence of a spiritual realm, a God or gods, a soul, and an afterlife. Strange that every culture should perceive reality with this same "spiritual" bent, that we should all hold such similar beliefs and then express them through such similar rites and practices. Are we to believe this is the result of some vast coincidence, or is it possible that we are compelled to maintain such beliefs as well as to engage in such practices as the result of a very sophisticated series of reflexes or instincts?

Similar to how all planarians have a tendency to orient themselves toward light, the fact that every human culture has a tendency to believe in a spiritual reality would imply one of three things. The first reason all cultures may have conceived of the same spiritual concepts would be as the result of some vast coincidence. This would be tantamount to believing that all planarians orient themselves toward the light for the same reason. Both possibilities are equally unlikely.

The second possible reason is that during the emergence of our species, the concepts of a spiritual realm, a god, a soul, and an afterlife were created by a few inspired individuals whose innovative ideas were verbally passed from one generation to the next as our species spread across the continents, disseminating these concepts around the globe to every culture. This would imply that our cross-cultural belief in a spiritual reality represents a learned as opposed to an inherited behavior.

The problem with this possibility is that, as our species spread across the globe, it's highly unlikely that any learned behaviors or beliefs could have made their way to every single community, no matter how remote, and then so tenaciously endure in each and every one of them. Learned–as opposed to inherited–behaviors come and go like the wind. This is why, for instance, though a multitude of languages have come and gone throughout our species' history, the impulse to create language has existed among every culture as a constant. The same, I am suggesting, is true for religious and spiritual belief. Though scores of spiritual belief systems (religions) have come and gone throughout our species' history, the spiritual/religious* impulse has persisted as a constant. Similarly, though scores of specific religious rites, practices, and beliefs have come and gone with time, the fundamental beliefs in a spiritual realm, spirits/supernatural beings/gods, a soul, and an afterlife have persisted throughout. These pivotal beliefs represent the foundation of every world religion. It is simply the manner in which these primary beliefs are manifest that is constantly shifting and evolving. The fact that these primary beliefs have so persistently endured among every culture and under such diverse environmental and historical circumstances leads me to believe that, just as in the case with language, there must exist some underlying physiological force at work here.

---

*I would like to make the important distinction between two separate human impulses: one of religiosity, the other of spirituality. The "religious" impulse compels us to engage in a variety of shared ritualistic behaviors such as church attendance and adherence to church codes and customs. This impulse therefore functions as a social adaptation, one that serves to provide us with a common set of mores, beliefs, values, and motivations, thereby reinforcing the group dynamic. As a social organism, it is necessary that we maintain a common ideology as it serves to sustain the survival strategy of strength in numbers–basic biophysics. Moreover, the religious impulse not only fosters the group dynamic, but also provides the individual with a necessary sense of purpose and community.

Unique from this, the "spiritual" impulse generates an altered state of consciousness (to be discussed in chapter nine), one that evokes feelings of awe, serenity, and ecstasy. Because we are "wired" to ascribe spiritual status to all things–including our own experiences–we tend to interpret these altered states as evidence of some divine or transcendental reality. As certain religious customs such as contemplation, chant, prayer, and engaging in church ritual can evoke a "spiritual" experience, the religious and spiritual impulses often work simultaneously to help bolster our faiths in a god as well as the church. Regardless, though these two impulses are integrally interrelated, they are, nevertheless, unique from one another and don't necessarily coincide. It is for this reason that it's possible for a person to be highly religious (devoted to church doctrine and ritual), though completely aspiritual (incapable of having a spiritual experience). Inversely, it is equally possible for someone to be highly spiritual, though not at all religious.

Take, for example, our feelings of grief or sadness. Why is it that all humans express these sentiments in the same way? Why is it that all humans cry? No one has to be taught to shed tears when mourning the death of a loved one. This is something we do innately, a reflex. But let's imagine for the moment that crying was a learned behavior. Imagine we had to be taught to cry as a means of expressing grief. If this were the case, wouldn't it be likely that at some point, some culture–just one–would have deviated from its original teaching and eventually come up with some other means of expressing this sentiment? If crying were a learned behavior, it is highly unlikely that every culture on Earth would, to this day, all express grief in the same exact manner. Analogously, the same principle can be applied to our universal spiritual beliefs and practices.

Assuming spirituality/religiosity is not learned, this leaves us with the last possibility: that our universal spiritual/religious proclivities represent an inherent characteristic of our species, a genetically inherited trait. This would mean that we're innately predisposed to believing in a spiritual reality. If true, we must then possess neurophysiological sites from which such spiritual perceptions, sensations, cognitions, and impulses are generated. Moreover, if we possess such sites in our brain, this further suggests that they emerge in us as the result of information stored within our genes, thus implying that humans possess what we could call "spiritual" genes.

Such a genetic interpretation would suggest that religious belief represents an inherent part of human nature and will emerge in any given society with the same determination as any of our other inherited instincts. The sociobiologist Robin Fox expressed this same notion in his work *The Cultural Animal.* While hypothesizing on the nature and development of a society of children reared in total isolation, Fox asserted:

> I do not doubt that they [the children] could speak and that, theoretically, given time, they or their offspring would invent and develop a language despite their never having been taught one. Furthermore,

> this language, although totally different than any known to us, would be analyzable to linguists on the same basis as other languages and translatable into all known languages. But I would push this further. If our new Adam and Eve could survive and breed–still in total isolation of any cultural influences–then eventually they would produce a society which would have laws about property, rules about incest and marriage, customs of taboo, a system of social status, courtship practices including the adornment of females, dancing, schizophrenia, homosexuality, initiation ceremonies for young men, myths and legends, and beliefs about the supernatural and practices relating to it.[25]

Imagine we were to study ten separate and totally isolated colonies of honeybees, all of which constructed their honeycombs in the same hexagonal pattern. After witnessing this, would we say that such behavior represents an example of "free thinking" bees all coincidentally building their hives in the exact same way? Or would we instead say that the bees, as a species, must be neurophysiologically "hardwired" to construct their hives in such a way–that is, that they do this as the result of a genetically inherited reflex? Do all bees construct their hives in the same hexagonal fashion because they willfully "choose" this design or because they are preprogrammed to build them in this particular way? In such a case, I imagine we would agree that bees construct their hives in this identical fashion as the result of a physiological impulse, that somewhere in the bee's brains there must exist a series of neural connections that compel them to construct hexagonally-shaped hives.

With this in mind, why, I ask, should we view our own universal (cross-cultural) behaviors any differently than we might the behaviors of bees? In the words of the founder of the science of sociobiology, E. O. Wilson, "The same principles of population biology and comparative zoology that have worked so well in explaining the rigid systems of the social insects could be applied point by point to vertebrate animals." If we are ever to make any progress in the understanding of our own physical natures, mustn't we study and assess ourselves with the same objectivity that we do all the Earth's other creatures? If a group of aliens were to study our species from above, what might they conjecture after witnessing approximately one hundred thousand years of the vast majority of our species ritualistically disposing of its dead in a hole in the ground? Would they not view such behavior as representative of an instinct? Would they not regard our behavior similar to the way we view the universality with which all planarians turn towards the light or all cats meow? Wouldn't these aliens surmise that the burial of the dead must represent an inherent characteristic of our species, the effect of a genetically inherited impulse or instinct?

In the same way that planarians are "hardwired" to turn towards the light, humankind is "hardwired" to turn to a god or gods. Being that this impulse is cognitive in nature, it must originate from a part or parts of the brain. Consequently, there must exist specific neural connections from which our spiritual/religious cognitions, perceptions, sensations, and behaviors are generated. This would further suggest that should we sever or alter these neural connections, these "spiritual" parts of our brain, it would have a direct affect on one's spiritual consciousness. For instance, should these parts of a person's brain be surgically altered, it's likely that individual would lose his sense of spiritual consciousness. Never again would that person have a spiritual experience. Never again would he feel the comforting presence of a protective spiritual force or entity. Never again would he feel compelled to pray, to look outwards to a transcendental force or being for guidance or assistance. Similar to the manner in which a person can develop a linguistic or musical aphasia, I'm suggesting that it's possible to develop a spiritual aphasia. For example, when a priest suffers from Alzheimer's, does he not lose, along with his other sensibilities, his sense of spiritual consciousness? Are we to believe that though this person can't feed or go to the bathroom by himself, he will still be able to pray or preach the gospel with lucidity? Apparently, spiritual consciousness is just as integrally linked to our neurophysiological makeup as is any of our other cognitive capacities.

Offering physical evidence to support this notion that humans can suffer from spiritual/religious aphasias, the Canadian psychologist Michael Persinger found that "one of the main differences between the 19 percent of high school students who had religious experiences before their teens, and the rest, was the presence of a head injury or blackout at least once during childhood."[26]

To further support Persinger's findings, Dr. Arnold Sadwin, as chief of neuropsychiatry at University of Pennsylvania's graduate hospital, came across people who had incurred religiously oriented personality disorders after incurring a blow to the head (what is known as an organic psycho-syndrome). In his research, Sadwin discovered individuals who, after suffering a head injury, showed distinct changes in their religious attitudes and behaviors. In some cases, he found individuals who, though they were extremely religious prior to their accidents, afterwards were indifferent to religious concerns. On the other hand, Dr. Sadwin also came across individuals who, though they were previously areligious, after experiencing a head injury, suddenly became hyperreligious, obsessively praying to God and expressing intense religious feelings and urges.

Most controversial of all, if such a genetic/neurophysiological hypothesis is correct, if the human species is "hardwired" to believe in a spirit world, this could suggest that God doesn't exist as something "out there," beyond and independent of us, but rather as the product of an inherited perception, the manifestation of an evolutionary adaptation that exists exclusively within the human brain. If true, this would imply that there is no actual spiritual reality, no God or gods, no soul, or afterlife. In such a light, spiritual concepts such as these would only exist as manifestations of the particular manner that our species has been "hardwired" to perceive reality. Consequently, humankind can no longer be viewed as a product of God but rather God must be viewed as a product of human cognition.

Just as Kant proposed that we inherit temporal and spatial consciousness, I'm suggesting we inherit our sense of spiritio-religious consciousness. Furthermore, just as Kant suggested that we are born with spatial and temporal modes of perception, two means through which our species is "wired" to interpret reality, I'm suggesting that spirituality represents yet another one of these inherent modes of perception. This would imply that our spiritual perceptions, like all others, aren't representative of any absolute truth but exist solely as a consequence of the manner our species is programmed to interpret reality.

Not only does this spiritual function act to transform our perception of reality, but it also seems to possess the ability to override our capacity for critical reasoning. This is made evident by the fact that though there is no physical evidence to confirm the existence of a spiritual reality, every culture has believed in one. This is rather unusual for a species of skeptics as ourselves. Generally speaking, human beings tend to believe only what their physical senses reveal to them. Unless we can see, feel, taste, smell, or touch something, we tend to be dubious of its existence. Nevertheless, our spiritual beliefs seem to represent an exception to this rule. Since there is no physical evidence to support the existence of any spiritual reality, it appears that our spiritual perceptions and beliefs must be originating not from information acquired from external sources via our physical senses but rather from information being generated from somewhere within.

For instance, should I tell the "average" person from any world culture that there were invisible pink elephants hovering about the room, chances are I would be ridiculed, if not subjected to a psychiatric evaluation. The reason my remark would prompt such a strong reaction would be that the information I would have conveyed would contradict everything that person's physical senses would reveal to him. Nevertheless, should I tell that same person that the spirit of God or of the deceased were hovering about the room, chances are he would be much more inclined to believe me, regardless of what his physical senses might convey. It would therefore appear that there exists some part of our brain that taints our perceptions and emotional responses in such a way as to compel us to sense supernatural forces all around us. That we possess such a cross-cultural proclivity suggests that we must be neurophysiologically hardwired this way.

To reiterate, if we apply the principle that all cross-cultural behaviors represent the effects of inherited impulses, it would suggest that human beings are genetically predisposed or hardwired to believe in the concepts of a spiritual reality, a God or gods, a soul, and an afterlife; to pray to and to worship these unseen forces; to ritualistically dispose of or bury the dead with expectations of an afterlife; to conduct religiously/spiritually oriented birth, initiation, marriage, and death rites; as well as to undergo "mystical" experiences. This would further imply that for every cross-cultural "spiritual" cognition, perception, or sensation we experience, there must exist some specific physical site or sites in the brain from which they are generated. Consequently, any damage incurred to these sites would alter or impair whatever specific "spiritual" perception, sensation, or cognition happens to be generated from that particular region. In summary, such a hypothesis suggests that all of our "spiritual" cognitions, perceptions, sensations, and behaviors are the manifestations of inherited impulses generated from neural connections in the brain and, therefore, not indicative of any actual spiritual reality.* But why, one might justifiably ask, if all cultures are instilled with the same inherent "spiritual" impulses, do so many different religions exist? Though we all possess the same regions in the brain from which our linguistic capacities are generated, each culture–based on its unique

---

*Though no one could ever prove that there is no spiritual reality, such a hypothesis certainly supports the possibility that one might not exist. As a matter of fact, it is impossible to prove that any imaginary force or being does not exist. How, for example, could one ever prove that there is no such thing as invisible pink elephants? Just because we've never seen one doesn't prove they don't exist. In this way, the mere act of trying to disprove the existence of a fantastical being is an exercise in futility. We must accept the principle that the burden of proof need lie in confirming something's existence, not its non-existence.

set of historical and environmental circumstances–develops its own linguistic identity or what we call a language.

Analogously, though we all possess the same regions in the brain from which our spiritual impulses are generated, each culture–based on its unique set of historical and environmental circumstances–develops its own spiritual identity or what we call a religion. By attaching sacred status to a unique set of people, places, objects, and customs, each society develops its own unique religion. Religion therefore represents the social medium through which our spiritual and religious impulses are given form and expression. The drive, therefore, to create a religion, with all of its codes, customs, and ritualistic behaviors, stands as its own distinct impulse.*

Similar to the manner in which all languages share the same fundamental rules of construction and syntax, each religion shares the same fundamental beliefs. Though each culture may believe in a different god, each believes in the existence of supernatural forces, in some form of a transcendental force or being. Though each culture may hold its own view of what death will bring, each believes in some form of an afterlife. Again, though we might all possess the same "spiritual" genes, the same "spiritual" function, because each culture has emerged from its own particular environmental and historical circumstance, each has developed its own unique mythology, its own religion.

This might help to explain, for instance, why more northern cultures, such as the Norse, incorporated such indigenous animals as bears, wolves, and whales into their religions, whereas desert-based peoples, such as the ancient Egyptians, incorporated animals as jackals, falcons, crocodiles, and snakes into theirs.

Presuming that spirituality represents the product of a genetically inherited impulse, I next had to ask: why would we have evolved such

---

*One of the key functions of the "religious" impulse is to regulate our drive to engage in repeated ritualistic acts. It is therefore possible that obsessive-compulsive disorders might constitute a dysfunction of this same impulse. In its healthiest form, the impulse to engage in repeated ritualistic behavior serves to reinforce our spiritual belief systems, promote social bonding, and give meaning and structure to our lives. In its dysfunctional form, however, we are instead compelled to compulsively engage in a repeated series of meaningless ritualistic acts and gestures.

a trait? What environmental pressure may have prompted the forces of evolution to select such a seemingly abstract trait as spiritual belief into our species? As all traits must serve to enhance a species' survivability, how might a spiritual function do this for ours? Furthermore, what is it about our species, in particular, that we alone should possess such a trait?*

Unless I could provide a sound explanation, a rationale, for why such a spiritual function might have evolved in us, it would be impossible to justify one's existence.

*With the exception of Neanderthal Man's simple bone altars and burial rites, no other species, besides ours, has given us any reason to believe that it might possess spiritual consciousness. Nevertheless, I have had others contend that this is a presumptuous assertion to make given that we can never really know what another animal is thinking. How can we know, for certain, that no other species senses a spiritual reality or believes in a god? Granted, though we can never "know" the thoughts of another species, based on their behavior, no other animal besides our own has given us any reason to believe that it possesses spiritual consciousness. When, for example, have dogs gathered around a ceremonial mound they erected and then bowed their heads in what might be suggestive of an act of deference or prayer? When has any chimpanzee carved or drawn a symbolic image of some imaginary or "spiritual" force or being? When has any other animal (besides the aforementioned Neanderthals, who were close phylogenetic cousins of ours) buried its dead in a ritualistic manner, suggestive of its conceiving some form of an afterlife? It is through an animal's behavior that we gain insight into the inner workings of its conscious experience, and none, other than our own, has given us any reason to believe that it possesses any semblance of spiritual consciousness.

CHAPTER 8

THE RATIONALE

"All that exists is rational."

–HEGEL

All that exists is rational. Every cause has its effect; every effect has its cause. In essence, nothing happens without a reason. Since this axiom applies to all that exists, it must also apply to all the various forms of terrestrial life–all forms including our own.

In applying this axiom to specific human characteristics, every trait we possess, from stereoscopic vision to our opposable thumbs, must have a specific reason for having emerged in us. Since the driving force behind all evolution is the preservation of a species, every trait must somehow serve to increase that species' chances of survival. This is evident in every organ we possess–excluding, of course, those vestigial parts such as the caudal vertebrae or coccyx (that evolutionary memento of our predecessors' tails) or the appendix (a relic of our grass-eating days), two examples of anatomical parts which, because we no longer need them, were selected out of us. Because all traits must perform a specific function that will serve to increase a species' survivability, if humans possess specific neurophysiological sites responsible for generating spiritual and religious consciousness, then the same must hold true for these parts as well.

We need therefore ask: What is the advantage of possessing spiritual consciousness? What function might such an adaptation serve that it could enhance our species' survivability? What is this trait's rationale, its reason for being? Again, as is true of all traits, if human spirituality didn't possess some very specific adaptive value, if it didn't somehow serve to enhance our species' survivability, it would never have emerged in us.

Most physical traits emerge in response to some environmental pressure. For instance, if Arctic wolves possess thick coats of fur, it's because their environments "pressured" them to evolve one. As our terrestrial environments are in a state of constant flux, organic matter–life–is constantly being forced to adapt to meet the demands of our ever changing conditions. Therefore, if humans do indeed possess a neurophysiologically based mechanism that compels us to believe in a spiritual reality, it's imperative we come to understand its purpose as well as its origins. If it's environmental pressure that forces the selection of new adaptations, then there must have existed some distinct environmental pressure that forced the selection of spiritual cognition upon our species.

In the case of the Arctic wolves, it was the pressure incurred by the cold weather that caused their thicker coats of fur to be selected. Among our own species, what environmental pressure might have prompted the evolution of a spiritual function on us? How might it have been to our advantage to believe in a spiritual reality, if, in fact, no such thing exists?

Moreover, what was so unique about our species that we alone should have developed such an unusual and abstract trait? Given that nature weeds out all that is superfluous, if spiritual consciousness did not somehow enhance our species' survivability, it simply would not have evolved in us.

The Origin of Mortal Consciousness

"In a hundred countries, in a thousand languages,
humanity stops and reaches upward, keenly
aware of its mortality."
–Peter Matthiessen

"No thought exists in me which death has not
carved with his chisel."
–Michelangelo

No other creature on Earth has the intellectual capacity of Homo sapiens. As a matter of fact, our intelligence constitutes the foundation of our species' remarkable strength. Whereas fish can swim, birds can fly, and cats have speed, humans possess an intelligence that has allowed us to venture deeper, fly higher, and move faster than any other creature on Earth. No other creature (besides the quasi-living viruses) comes close to challenging our dominion over the other forms of life. All we have to do is look around us to behold the awesome power of our intelligence. In the last hundred years alone, we have transformed our planet's surface more dramatically than any other species has in the last three billion.

Nevertheless, as much as our vast intelligence may have graced our species, it has also been the source of our greatest affliction. Though our intelligence may have made us the most versatile and therefore powerful creature on Earth, this same adaptation has backfired on our species with nearly the same potency that it has served us. As a result of our intelligence, something happened that had never before occurred within the known universe. With the same powers of perception that had allowed our predecessors to scrutinize the world around them, Homo sapiens developed the unique capacity to perceive their own selves. For the first time in the history of life, an organic form emerged that was aware of its own existence. No other creature before us had any idea, for instance, that when it drank from the watering hole, the image it gazed down upon was that of its own reflection. Now, for the first time in life's three-and-a-half-billion-year history, an organism–ours–suddenly could. For the first time in the history of the known universe, a combination of molecules had emerged that could comprehend its own existence.

Imagine those first primal humans looking down at their own hands, their own bodies, in awe of what they saw and, for the first time in terrestrial life's history, asking that fateful question, "What is this that I am? What is this that I exist?" With the capacity for this one cognition, this one self-reflection, the human species was transformed. In biblical terms, man had taken his first bite of the forbidden fruit fresh from the tree of knowledge.

It was probably not long after this first cognitive lightning flash that we were hit with the inevitable thunder: "If I am, if I exist, then isn't it conceivable that one day I might not?" With the same capacity with which humans could comprehend their own existence, we simultaneously became equally aware of the possibility of our own nonexistence...of death. With this one awareness, the wheels of life which had been turning so smoothly for all these billions of years had turned down a cognitive cul-de-sac. Humankind had suffered life's first existential crisis.

The Pain Function

"Pain and death are a part of life.
To reject them is to reject life itself."
–Havelock Ellis

According to Buddha, enlightenment can be attained by anyone willing to follow the path of the "Fourfold Truths." The first of these truths, which Buddha referred to as Dukkha, asserts that life is a process of universal misery and suffering. No matter who we are, be it prince or pauper, we are all destined to experience the same fateful demise. We are all bound to grow old, weak, and infirm. We are all preordained to lose everything we ever had or loved, including our own selves. In a nutshell, we are all doomed to die. Borrowing from this tenet of Buddhist pessimism, Freud expressed a similar notion:

> We are threatened with suffering from three directions: from our own body, which is doomed to decay and dissolution and which cannot even do without pain and anxiety as warning signals; from the external world, which may rage against us with overwhelming and merciless forces of destruction; and finally from our relations to other men.[27]

Because our lives are incessantly threatened by such perilous forces, pain represents not only a biological phenomenon but a biological necessity. Just as with every other trait we possess, we experience pain because it serves a very specific function.

But what exactly is pain? Pain is a negative sensation experienced by organic forms when specific receptors are triggered in the brain. Stimuli that elicit pain are generally indicative of things that represent potential threats to an organism's existence. For example, excessive heat can harm, if not kill, a creature. It is for this reason that many animals possess heat-sensitive receptors that cover the surface of their skin. When these receptors come in contact with excessive heat, an animal experiences this potentially hazardous stimulus as a negative sensation we call pain. By experiencing excessive heat in such a negative or "painful" manner, animals are compelled to avoid that which can burn them. Should an animal, for instance, get too close to a flame, the negative sensation of pain will prompt it to recoil, thus saving it from what may have caused more serious, if not irreparable, damage. Pain therefore represents an evolutionary adaptation meant to encourage organic forms to avoid those things that can threaten their existence. It is this pain function that keeps us ever-vigilant and prevents us from allowing ourselves to be cut or pierced, to burn, freeze, starve, or dehydrate.

To provide a specific example of how this pain function operates, I'll use the example of hunger in a rabbit. In order to prevent a rabbit from starving to death, its undernourished body will send a distress signal to its control center–its brain (specifically to the brain's thalamus from which the experience of pain is generated)–that it's in need of sustenance. It is this negative sensation that will motivate the rabbit to seek its required fuel supply. If this physical need is not met within a certain time frame, the animal's body will reinforce this signal by stimulating even more pain receptors, causing the rabbit's hunger to be intensified. What was previously experienced as a mild discomfort becomes acute pain. In essence, the body is sending a distress signal to itself saying, "Feed me or die!" In order to relieve itself of the painful sensation of hunger, the animal is motivated to seek out sustenance–to eat. Let's now suppose the rabbit finds itself some fuel or what we call food. In our own inaccurate language, when the rabbit finally consumes its meal, we tend to say that it is experiencing pleasure. If, however, we look at this from a purely biological perspective, it is not pleasure that the animal is experiencing but rather the diminishment of its discomfort or pain.

Just as the experience of pain increases an individual animal's survivability, it plays an equally important role in maintaining the preservation of a species. For example, it is the negative stimulus of sexual tension that incites all animals to reproduce. Among mammals, reproduction represents a hindrance to individual survival, as having to provide for offspring means an animal has that much less time to devote to securing its own personal needs. Giving birth to and rearing young therefore represent an obstacle to individual survival. Nevertheless, as reproduction plays such an integral role in the preservation of any species, it is a necessity. It is for this reason that all animals are biochemically driven to engage in sexual intercourse. Among humans, sexual deprivation incurs physical as well as psychological tension and discomfort (and among men can even increase the chances of incurring testicular cancer). Consequently, sexual release relieves one of normal sexual tension, illustrating that though sex is generally perceived as something pleasurable, it more accurately represents the diminishment of pain.

Among the "higher" order social animals, most particularly among Homo sapiens, another example of a negative or painful stimulus that serves to promote the well-being of the species involves that negative experience we refer to as loneliness. When one is alone, isolated from the community, he is most vulnerable. As no individual is completely self-sufficient, each of us must rely on the assistance, care, and protection of others. On our own, we are most defenseless. Within the group, however, an individual gains the added security and strength that comes with increased numbers. It is for this reason that nature selected a negative or painful stimulus we call loneliness that prompts individuals to pursue the company of others.

Another negative stimulus that serves to promote the well-being of the individual as well as its species involves what we refer to as "separation anxiety," a physical discomfort experienced when we are separated from a loved one. Because romantic love fosters procreation, security, and effective child-rearing, it is necessary that we experience discomfort when separated from our romantic partners. Consequently, though we perceive ourselves as joyous when reunited with a loved one from whom we've been separated, it is really the diminishment of our separation anxiety that we are experiencing.*

In summary, it is pain that keeps organic forms alive and intact. Pain is nature's electric prod that incessantly goads us towards those things which benefit us and away from those which can do us harm. We therefore experience pain and discomfort for a reason. Pain represents the chief stimulus by which all life is prompted to survive.

---

*A research team led by anthropologist Helen Fisher of Rutgers University has been working to determine the neurochemistry involved in bonding behaviors. Fisher believes that attachments formed between individuals "in love" are caused by changes in the brain involving a group of neurotransmitters called mono-amines, which include dopamine, norepinephrine, and serotonin. To plot these changes, Fisher subjected lovelorn couples to a functional magnetic resonance imaging (fMRI) brain scanner that could pinpoint minute changes in blood flow in the brain associated with bonding and infatuation. What she found was that whereas lust is governed by testosterone and estrogen, attachment is governed by the neurotransmitters oxytocin and vasopressin. Apparently, even romantic love and attachment can be reduced to neurochemical processes. This hypothesis was later confirmed when Andreas Bartles at University College London found that when students placed in an fMRI were shown photographs of loved ones (versus photos of insignificant others, which had much less effect), specific regions of the brain became highly activated. The areas which lit up were part of the anterior cingulate cortex, the middle insula, and parts of the putamen and caudate nucleus.

The Anxiety Function

"Just as courage imperils life, fear protects it."
–LEONARDO DA VINCI

"There are times when fear is good.
There is advantage in the wisdom won from pain."
–AESCHYLUS

Among the "higher" order animals, most particularly the mammals, threatening circumstances elicit a particular type of pain we call anxiety. Anxiety constitutes a specific kind of painful response meant to prompt these higher order animals to avoid potentially hazardous circumstances.

As the stomach is the organ responsible for the digestion of food, its pain receptors respond to the quality of nourishment it receives. Analogously, as the brain is where all data is stored, it is responsive to the quality of information it receives. For example, a baby rabbit pokes its nose into a fire for the very first time. The excessive temperature stimulates heat receptors dispersed throughout the rabbit's skin. This negative (painful) stimulus excites the motor reflexes which prompt the rabbit to recoil from the flame. Having escaped the situation with little more than a superficial burn, the rabbit will now encode this painful experience in the form of a memory. From now on, whenever the rabbit perceives a fiery object, the memory of its encoded experience will be retrieved, thereby alerting it not to repeat its past action. Rather than having to experience being burned over and over again, the rabbit's memory will now act as a buffer against all possible future experiences with objects that emit excessive heat.

Though this capacity to store and utilize memories enables the rabbit to avoid fire without having to be burned over and over again, this does not mean that the memory itself is altogether pain-free. In order to remind the rabbit of the potential threat that fire and excessive heat represent, the memory will elicit a type of discomfort we call anxiety. In this way, though anxiety may serve to protect the rabbit from incurring any actual physical injury, it nevertheless evokes a certain degree of discomfort. That an actual memory can cause one to experience psychological discomfort (anxiety) demonstrates that memories store emotional as well as purely perceptual data. As a matter of fact, emotional memory has been attributed to the brain's amygdala, which, when damaged, can result in the loss of an individual's capacity to retrieve memories that contain emotional content (Le Doux, 1994).

With this advanced faculty to store emotional memories, in conjunction with the capacity to experience anxiety, an organism no longer had to sustain actual physical injury before it was motivated to avoid a potentially hazardous experience. Anxiety therefore acts as an early warning device that keeps an organism ever alert to potential threats before one is actualized.

In another, more extreme example of how anxiety serves us, imagine that the rabbit now crawls into a cave to find itself suddenly face-to-face with a fierce mountain lion. The perilous nature of the situation causes the rabbit to experience the most painful symptoms of anxiety, all meant to compel it to escape its potentially hazardous circumstance. Some of the negative symptoms of anxiety include heart palpitations, muscle tension, hyperventilating, trembling, perspiration–all which are meant to prompt the rabbit to get as far away from the source of its discomfort (in this case the mountain lion) as quickly and effectively as possible. Consequently, even though the mountain lion has yet to lay a paw on the rabbit, the rabbit will still experience the pain of its own anxieties.

In a case in which an animal is confronted by such a mortal threat as this, the symptoms of anxiety can be extremely painful. Anxiety therefore serves as an advantageous adaptation in that it prompts an animal to respond to a potentially hazardous situation with greater speed and efficiency. Should our rabbit manage to escape the mountain lion, it will encode this anxiety-engendering experience in the form of a memory. Now, the next time the rabbit leaves its lair, the anxiety-evoking memory of its past experience with a mountain lion will discourage it from going anywhere near one. Thanks to this anxiety function, our rabbit no longer needs to be attacked by a mountain lion over and over again to know to avoid one. It is for this reason that anxiety represents a biological necessity. As Ernest Becker, author of *Denial of Death*, wrote:

> Animals, in order to survive, have to be protected by fear responses, in relation not only to other animals but to nature itself. They had to see the real relationship of their limited powers to the dangerous world in which they were immersed. Reality and fear go together naturally.[28]

As the human brain is more complex than that of all the other species, our cognitive capacities are that much more sophisticated. First of all, our brains contain much more storage space, enabling us to retain many more memories. Secondly, our species possesses an enhanced capacity to comprehend our own possible futures. As a result of the combined effects of these two capabilities–because humans are aware, for example, that hunger elicits pain, enhanced by our capacity for foresight–we are motivated to procure food and shelter not just for today but for our futures. Unlike many of our evolutionary ancestors who needed to rely on the immediate stimulus of hunger to be motivated to search for nourishment, human beings are compelled to make sure there is food available long before it is actually needed. This capacity for foresight grants us the added benefit of having more time to secure our most basic vital needs. Because a simpler organism needs to rely on the immediate stimulus of hunger to be prompted to search for its needed food supply, it may only have a few days' advance notice to procure its next meal before it will starve. In the case of humans, however, as a result of our advanced capacity for foresight, we are compelled to search for food long before we are even hungry.

Though this capacity for foresight may work to our advantage, it comes with a serious drawback. Due to our incredible capacity for foresight, instead of just being anxious about those threats that exist in the present, humans experience anxiety for all those possible threats that might jeopardize us in the future. Consequently, humans don't just experience anxiety over how they will procure their next meal but over how they will secure tomorrow's meals as well. And it's not just tomorrow's meals we're concerned with, but all those we will ultimately need to sustain ourselves way into the future–if not for the rest of our lives. For this reason, though our capacity for foresight may serve to our advantage, it at the same time engenders a tremendous amount of anxiety.

In many ways, the anxiety function represents our primary defense in our incessant struggle for survival. It is this anxiety function that keeps us ever vigilant and alert, always on guard against the potential threats of hunger, dehydration, excessive heat or cold, strangers, disease, predatory animals, poisonous plants, sharp objects, fires, floods, droughts, hurricanes, the dark, etc., all things we have the unique ability to secure ourselves against long before they represent an actual threat. It is this anxiety function that has motivated us to manufacture fire and electric light, to develop all sorts of medical technologies, to build dams and structural fortifications, to erect silos to store vast deposits of food, and to devise methods of refrigeration. Due to our enhanced capacity for foresight combined with the anxiety induced by our fear of potential future threats, we are obsessed with our futures. It's necessary we be this way, for the minute we become lax and lower our guards, we become vulnerable to a world of potential hazards and predators. In essence, the less anxious we are, the more vulnerable and therefore endangered we become.

Whereas other animals may have claws or sharp teeth with which to protect themselves, humans possess a capacity for foresight. With our enhanced capacity to envision our possible futures, humankind is that much more equipped to fortify itself against more threats than any other creature. Nevertheless, this type of advanced intelligence comes at a very high price.

When Mortal Consciousness Meets the Anxiety Function

"Anxiety is the state in which a being is aware of its possible non-being...The anxiety of death is the most basic, most universal and inescapable."[29]
–PAUL TILLICH

"No one is free from the fear of death...The fear of death is always present in our mental functioning."[30]
–G. ZILBOORG

"The deep realization of the frailty and impermanence of man as a biological creature is accompanied by an agonizing existential crisis."
–STANISLAV GROF

"He that cuts off twenty years of life cuts off so many years of fearing death."
–SHAKESPEARE, *JULIUS CAESAR*, ACT III

So what becomes of our anxiety function when it is confronted by our species' unique awareness of death? How are we to effectively utilize our capacity for foresight when it is incessantly informing us that we are ultimately going to die?

It is our capacity for foresight complemented by our anxiety function that keeps us perpetually vigilant, always on the lookout for any potentially hazardous situation. And though it is this same awareness that motivates us to avoid such perils, it, at the same time, brings us face to face with the fact that no matter what we do to fortify ourselves, our actions are all in vain. No matter how hard we work to provide ourselves with food and shelter, no matter what we do to protect and defend ourselves, no matter how much we plan and prepare for our futures, we know that death is inevitable and inescapable. It is this awareness that strips the anxiety function of all its efficacy, in turn, stripping humankind of its capacity to effectively survive.

No other creature on this planet can comprehend the concept of its own existence. Consequently, no other creature can conceive of its own nonexistence, of its own mortality, of death. This coincides with the fact that no other creature can comprehend the concept of its own future. Before us, all creatures lived in and for the moment. If an animal got hungry, it sought food. If it got tired, it slept. It lived and it died without one conscious thought regarding its own mortal existence or nonexistence. It had no conceptual awareness of its own possible future and therefore of its own possible death. The question "What might happen to me tomorrow?" had never before been asked until man conceived that such a day existed. As aptly phrased in the *Encyclopædia Britanica*, "This time consciousness, which is possessed by no other species with such insistent clarity, enables man to draw upon past experience in the present and to plan for future contingencies. This faculty, however, has another effect: it causes man to be aware that he is subject to a process that brings change, aging, decay, and ultimately death to all living things. Man, thus, knows what no other animal apparently knows about itself, namely that he is mortal. He can project himself mentally into the future and anticipate his own decease. Man's burial customs grimly attest to his preoccupation with death from the very dawn of human culture in the Paleolithic age. Significantly, the burial of the dead is practiced by no other species. The menace of death is thus inextricably bound up with man's consciousness of time."[31]

To add insult to injury, not only are we aware that we must die, but we also know that death can come at any given moment. Regarding our futures, nothing is certain. We live our lives anxiously standing beneath the mythical sword of Damocles, awaiting the day when that single strand of hair that holds inevitable death suspended precariously above our heads will finally snap.

Imagine how apparent this must have been to our earliest ancestors. How much security did primitive humans have that each day would not be their last? Imagine a time when there was hardly any real knowledge of medical science, when what may have seemed like an innocuous belly- or toothache one day brought death the next. What constant dread and uncertainty must have plagued our ancestors' existences. Among such nomadic dwellers, even the seemingly simple task of procuring one's next meal represented a potentially mortal chore. Whereas today we can merely pull up to the nearest drive-through restaurant to obtain our daily ration of meat, these men had to go out with their crude hunting utensils and bludgeon some ferocious beast to death in order to procure their next meal. In such times, the threat of death was constant. And yet, with all of our modern conveniences and medical technologies, very little has really changed. Even with all of our advancements, there is still no escaping the fact that we are all destined to die and that death can occur at any moment. Sure we may live another twenty or thirty years longer than our predecessors, but what difference does that really make when measured against eternity?

Living with certain knowledge of imminent death leaves us in a perpetual state of anxiety. At every moment, we stand metaphorically face-to-face with a mountain lion from which there is no escape, staring straight into the jaws of death. Consequently, we are forced to live out our existences in a state of unrelenting mortal terror and dread.

The chief difference between our condition and that of the rabbit as it stands face to face with a mountain lion is that whereas the rabbit can escape the object of its fear, human beings cannot. Since we became cognizant of inevitable death, we have been in a state of unremitting mortal fear of an enemy we cannot see, flee, or defeat. In essence, we are no better off than if we were born with a time bomb strapped to us set on a random timer to explode at any given moment within the next fifty or so odd years. What would we do in such a case other than to spend the rest of our lives in a state of constant peril and dread, waiting for the ticking time bomb to finally detonate? How, I ask, is the human condition any different from this? The threat of death lurks around every corner, in every breath, shadow, meal, and stranger. And though we don't know from where it will come, we are condemned to recognize that it inevitably will.

In addition to this, almost as potent as our fear of personal death is the fear of losing those we love. As a social organism, we are dependent on others for our physical as well as emotional survival. Again and again, studies show the debilitating effects of isolation in humans. Without love, we are generally pained beings.* For this reason, we place nearly the same–if not more–value on the lives of those to whom we are emotionally attached as we do on our own. Consequently, we live in constant fear not just of losing our own lives but of losing the lives of those we cherish and love.

Just as there is no escape from death, there is no escape from the consequent anxiety that our mortal awareness imposes upon us. With the advent of our awareness of death, humankind was left in a state of perpetual angst or what Kierkegaard called "the sickness unto death." With the dawn of self-conscious awareness, the anxiety function had imploded, rendering us a debilitated and ineffectual organism.

It is this breakdown of our anxiety function that makes human beings the dysfunctional animals we are. In our frivolous attempts to either oppose or escape unavoidable death, we channel our energies into a morbid array of self-destructive behaviors. In our futile efforts to oppose the unopposable, we have become the only animal that will needlessly kill one another as well as our own selves. Unlike any other creature on Earth, we are capable of acts of suicide, genocide, sadism, masochism, self-mutilation, and drug abuse, along with a multitude of other disturbed responses, all of which result from our species' unique capacity for self-conscious

*This was most effectively demonstrated by the pioneer work of Harry Harlow, who raised infant monkeys in varying degrees of isolation and found that those reared without ample maternal love developed a host of neuroses. In the most extreme example, those reared in solitary confinement grew to be utterly dysfunctional adults who, to compensate for their lack of any contact, spent their days crouched in a corner, trembling in fear and chewing on their own limbs as a means to provide themselves with needed sensual stimulation.

awareness and with it an awareness of death. As a result of our advanced capacity to conceptualize our own deaths, humankind had become a psychologically unstable entity, or as Freud phrased it, the "neurotic" animal.

Furthermore, in light of our awareness of inevitable death, life takes on a newfound sense of existential meaninglessness. Our struggles to survive become an exercise in futility. Between death's inevitability and all of the suffering we are forced to endure while awaiting our demise, we are compelled to ask: "Why go on living? What's the point?" How was our species to justify its continued existence in light of such a hopeless and desperate circumstance? Why struggle today when tomorrow we won't even be here? Under such conditions, the motivating principle of self-preservation that had sustained life for all these billions of years no longer applied to our species. This was a whole new set of rules our animal was now playing by, and unless something could be done to ameliorate our species' pained and desperate circumstance, it might not have been long before our newly evolved animal would have succumbed to the forces of extinction.

Advent of the Spiritual Function

"Fear begets gods."
–LUCRETIUS

"In order to counter this fundamental angst, humans are 'wired' for God."[32]
–HERBERT BENSON

"If the brain evolved by natural selection...religious beliefs must have arisen by the same mechanism."
–E. O. WILSON

So there we were, a newly emergent species with an unparalleled intelligence, one that had made us the most powerful creature on Earth. And then, as everything seemed to be working just fine, the inevitable took place: man's intelligence backfired on him. For the first time in the history of life, an organic form turned its powers of perception back upon its own self, rendering it aware of its own existence. With the dawn of self-conscious awareness, a cognitive revolution had taken place. With a newfound awareness of its own existence, the human animal had become equally aware of the possibility of its own non-existence. And so, with this one cognition, the most powerful creature on Earth was suddenly incapacitated by a crippling awareness of its own inevitable death.

Imagine how these first protohumans must have felt, suddenly cognizant of their own inevitable demise–naked, vulnerable, alone, defenseless against the threat of impending death, exposed before the void, unprotected by any "higher" force or being. If nature didn't provide our newly emergent animal with some type of adaptation through which to counter the anxiety induced by mortal awareness, it's quite possible our species might not have endured. In order to compensate for this debilitating awareness, nature was going to have to modify our animal's cognitive processing in such a way that we would be able to survive our unique awareness of death.

Rather than being stricken by some devastating new viral or climatic threat, humankind was now being assailed by an environmental pressure that just so happened to originate from within our own heads (after all, don't our own bodies constitute our physical environments?). As a result of this new internal and physiologically based environmental pressure, it became necessary that hominid cognition continue to be transformed if the line was to survive.

In response to this new environmental pressure, the forces of selection could have affected our evolution in one of two ways. Essentially, our intelligence, which had served as our greatest strength, was now jeopardizing our very existence. One evolutionary strategy that "nature" could have employed would have been to weed out the more self-aware members of our species, thereby leaving a population of less mortally conscious individuals to survive. In other words, the forces of natural selection could have simply pushed us back a few stages in our cognitive evolutions and returned us to our former, less self-consciously aware, less intelligent states. The problem with this solution, however, is that self-conscious awareness represents one of our species' most formidable capacities. Because we are self-aware, we possess the unique capacity to adapt ourselves to any situation or environment. For example, should we be faced with another ice age, whereas any other animal would have to wait millions of years for nature to select a thicker coat of fur, humans can sew themselves one within a few hours' time. As a result of our vast intelligence, Homo sapiens have outgrown the forces of evolution. We no longer need to wait for natural selection to alter us as we possess the unique capacity to alter ourselves to suit almost any physical environment. As a result of our incredible capacity to transform our immediate environments, humans now have the capacity to survive everything from the ocean's depths to outer space. Because we possess such advanced intelligences as for language and mathematics, humans can create tools and technologies that enable us to overcome almost any physical shortcoming. Environmental pressures that might wipe out another species simply push humankind to technologically progress, enabling us to adapt to our environments without the aid or benefit of natural selection.

At the same time, compromise our intelligence and human beings constitute one of the weakest and most vulnerable creatures on Earth. Devoid of such defensive adaptations as claws, fangs, wings, or venomous sting or discharge, without our intelligence man is like a walking meal waiting to be eaten. Consequently, the weeding out of intelligence probably wouldn't have been the most effective strategy. Rather, nature would be forced to select some new adaptation if humankind was to survive mortal consciousness. What kind of adaptation could accomplish this? What mechanism could possibly emerge in us that would relieve us of our incapacitating awareness of death without compromising our intellectual faculties?

Perhaps, at first, only those individuals whose cerebral constitutions somehow withstood the crippling anxiety that came with self-conscious/mortal awareness managed to survive. Nevertheless, something more was needed if the species, as a whole, was going to endure. Perhaps humankind's newly emergent awareness of death created so much tension in our animal that it induced a selective pressure on our cerebral physiologies. Just as environmental pressures transform entire species, why shouldn't these same pressures be able to transform our organ, the brain? Shouldn't those same Darwinian principles that apply to all organic matter apply to our cerebral evolutions as well? How else are we to imagine that all of our other cognitive centers–be they linguistic, musical, or mathematical–emerged?

As a result of our species' capacity for self-conscious awareness, we suddenly needed to be reconfigured in such a way that we could meet the new demands imposed on us by our internal environments. What this meant was that those individuals whose brains possessed some genetic mutation that could withstand the overwhelming anxiety induced by our awareness of death were more likely to survive. Those more likely to survive, consequently, were more likely to pass whatever advantageous adaptation they possessed onto their offspring.

As generations of these protohumans passed, those whose cerebral constitutions most effectively dealt with the anxiety resulting from their awareness of death were most apt to survive. This process continued until a cognitive function emerged that altered the way these protohumans perceived reality by adding a "spiritual" component to their perspectives. Just as the human brain had evolved linguistic, musical, and mathematical intelligence, we apparently evolved "spiritual" intelligence as well.

In summary, our species' awareness of inevitable death placed such a strong pressure on our cerebral (cognitive) evolutions that at some point during the latter stages of hominid evolution, nature selected those lineages which possessed a built-in predisposition to believe in or perceive an alternate reality, one that supersedes the limitations of this finite physical realm which can only offer us pain, suffering, and ultimately death. And so, a new reality was born in man, one which compelled our species to think to believe itself transcendent, to imagine that we are *more*, perhaps, than we actually are.

The Origins of Immortal and God Consciousness

Of those factors that may have influenced the evolution of a "spiritual" cognitive function, one, I believe, to have played a key role incorporates man's unique capacity to enumerate. Most animals possess an innate comprehension of the dimensions of time and space. Because we live through time and in space, it is necessary that we possess such an inherent awareness in order to survive. For instance, most animals possess an internal biological clock, one that serves to regulate an organism's behavior in relation to time. This biological clock will regulate what time of the day or year an animal will forage, sleep, or mate, as a few examples.

Many animals rely, to a large extent, on their sense of sight for survival. Because our planet's lighting conditions are determined by the Earth's rotation around the sun, this orbital cycle plays a critical role in most animal behavior. Furthermore, because our planet's revolution around the sun plays a critical role in the Earth's climate, this, too, will have a dramatic effect on a great deal of organic behavior. Because our environmental conditions are framed by time, it's necessary that most animals possess an internalized biological clock that can help them to effectively utilize the Earth's cycles of climate and light.

Besides possessing an inherent perception of temporal events, all life forms possess a built-in mechanism that enables them to perceive the world spatially. Even a plant, though it may be rooted to the ground, engages in the heliotropic propensity to turn its leaves towards the sun. Because we exist within a three-dimensional (spatial) environment, most animals possess some combination of organs through which they can discern up and down, backward and forward, near and far. As mobile creatures, it would be impossible for an animal to survive without such spatial sensibilities.

Though most animals possess a certain degree of temporal and spatial awareness, our species' capacity to comprehend both of these dimensions is by far the most advanced. Only humans can discern increments of time and space with such precision. By being able to apportion our world into such discrete spatial and temporal units, humans have evolved the capacity to enumerate objects–to count.*

Because our species possesses this particular "mathematical" cognitive capacity, humans are able to measure moments in time as well as units in space. Consequently, as a result of possessing such an enumerating or mathematical function, we alone have been able to navigate our way across the oceans and seas, the continents, and, most recently, extraterrestrial space. This capacity has also enabled us to construct immense architectural fortifications, countless machines and technologies, along with formidable instruments of healing as well as destruction, all things which, for better or worse, have served to make us the most powerful creature on Earth.

Although this capacity has generally worked to our advantage, just as in the case of self-conscious awareness, our ability to enumerate affected us in a similarly hazardous way. The reason for this is that inherent in our capacity to enumerate–to add one plus one–exists the intrinsic realization that this process has no finite end (i.e., no matter how big a number is, we can always add one to it). Consequently, as a result of our advanced ability to enumerate exists an intrinsic capacity to conceptualize infinity. As only our species possesses this sophisticated a capacity to enumerate, only we have this capacity to comprehend the concept of infinity.**

*It was recently discovered that Rhesus monkeys possess the capacity to enumerate objects in consecutive order from one to nine. Here is an example of a closely phylogenetically related ancestor possessing an incipient talent for a predominantly human capacity.

**As mathematical consciousness represents a cross-cultural characteristic of our species, this would suggest that mathematical ability must constitute a genetically inherited trait. This would further imply that there must exist "mathematical" sites within the brain. The existence, for instance, of mathematical idiot savants, people who can calculate into the billions but who are otherwise cognitively impaired, would seem to confirm the existence of such a neurophysiological mechanism. Moreover, as every culture has–either through words or symbolic images–conceptualized infinity, this would further imply that there might exist a specific part of our mathematical function that enables us to conceive of this particular abstraction. Moreover, if such a neurophysiologically based "infinity" site within our brain does indeed exist, it then follows that we must also possess what we could call "infinity" genes responsible for the emergence of those sites.

In the same way that we can enumerate units in space, we can do the same with moments in time. And just as we can comprehend the idea that one plus one equals two, we can equally conceptualize the notion that this present day plus one more equals tomorrow. It is from this same cognitive faculty that humans may have gained their capacity for future consciousness or foresight, one that has enabled every human culture to devise a calendar by which it can measure the foreseeable future in days, seasons, and years.

Just as our enumerating capacity has enabled us to conceptualize that spatial dimensions possess no finite end, we can equally apply this same notion to temporal dimensions. Analogous to the way we can conceptualize infinity, we can equally conceptualize eternity. Just as we can keep adding one unit to any spatial dimension, ad infinitum, we can do the same with temporal dimensions as well (this moment plus the next moment equals the moment after, and so on and so on, ad infinitum). With this capacity to conceive that temporal dimensions have no finite end, not only can we conceptualize our own futures all the way to our inevitable deaths but way beyond that into eternity. Because we can comprehend the concept of eternity, our species must live with an awareness that though we, our physical selves, are temporal in nature, time itself will never end. With a conscious awareness of eternity, humans were suddenly forced to endure the notion of how infinitely brief life is. Whereas all other creatures live in and for the moment, we now had to measure our existences against the overwhelming backdrop of all eternity. Suddenly, humankind had to contend with an inherent sense of its own ultimate and painful insignificance. In the words of the philosopher Blaise Pascal, "the finite is annihilated by the infinite." Consequently, due to our capacity to grasp the eternal and the infinite, our species now had to endure a new anxiety, one which may have rivaled that which came as a result of our debilitating awareness of death.

Due to our capacity to comprehend the infinite and eternal, it seems that mathematical intelligence may have played just as significant a role in the evolution of a spiritual function as did our awareness of death. Not only did we now need to be protected from actual death itself but from all of the possibilities that might exist long after death. Suddenly, man was aware that he might exist (or for that matter, not exist) for all eternity. But how? In what form? Would eternity be a pleasurable or a painful experience? Would we retain our conscious identities and, if so, in what state? Would life after death be as replete with experience as this life or would it represent a state of absolute nothingness, of eternal nonexistence? Furthermore, what might that even mean? As it is natural for our animal to be concerned with our futures, humans were suddenly condemned to spend their lives no longer just in fear of death, but in fear of what might come after death, in fear of the possibility of eternal suffering or, perhaps even more disconcerting, of eternal non-existence.

Rather than allowing these fears to overwhelm and destroy us, perhaps nature selected those whose cognitive sensibilities compelled them to process their concept of death in an entirely new fashion. Perhaps after hundreds of generations of natural selection, a group of humans emerged who perceived infinity and eternity as an inextricable part of self-consciousness and self-identity. Perhaps a series of neurological connections emerged in our species that compelled us to perceive ourselves as "spiritually" eternal. Once we perceived ourselves as possessing an element of the infinite and eternal within us, as apparent as it was that our physical bodies would one day perish, we were now "wired" to believe that our conscious self, what we came to refer to as our spirit or soul, would persist forever. As a result humans began to view themselves as immortal, a concept that has endured universally among nearly every single culture from the dawn of the species.*

---

*Mathematical or numerical consciousness is apparently integrally interrelated with our sense of spiritual consciousness. This relationship is made evident by the fact that every world culture has attributed spiritual significance to numbers and geometric shapes. Whether it be the Jewish Kabbalists, the Pythagorean Greeks, the medieval alchemists, the Christian use of a holy trinity, the use of numbers in Aztec mythology, numerical references made in the I Ching, or the general use of numbers employed by a variety of astrological and numerological belief systems, every world culture has maintained a belief that numbers can possess sacred power and significance.

Herein lies the cognitive origins of our cross-cultural belief in immortality, in our inherent perception that we–by virtue of our eternal souls–transcend physical death. Once we came to perceive consciousness as eternal in nature, we perceived physical death as nothing more than just another life-passage in eternal existence. Suddenly our animal was compelled to bury its dead with a rite that anticipated sending the deceased's eternal self or "soul" to another realm, or what developed to become an inherent belief in an afterlife. With the advent of this inherent inclination to believe in immortal existence, our species was relieved of a large part of the anxiety induced by our fear of imminent and eternal death. Humankind was saved.

But even if we were to live forever, what did that mean? Humankind still needed relief from the fear of the unknown. Would the afterlife be a place of eternal peace and happiness? Or would it perhaps be even more painful and precarious than our stay here on Earth? Without our parents to protect us in the afterlife, humankind now needed eternal guidance and protection from all that might come in the hereafter.

According to Freud, "God is the exalted father, and the longing for the father is the root of all religion."[33] Aware that death was not only inevitable, but that it could come at any moment, human beings were reduced to a state of infantile helplessness, as vulnerable as the day they were born. And where do infants innately turn for protection? To their parents. However, not even one's parents can save one from death. As we become adults, we grow to recognize that even our once seemingly omnipotent parents are actually impotent against the forces of death. With this knowledge, where was humankind to find guidance and protection? Desperately longing for eternal comfort and security, to whom or what was primal man to turn? Perhaps our need for eternal protection had facilitated the selection of a cognitive variation that instilled our species with an inherent belief in some type of a transcendental guardian. Perhaps it was at this point in human cognitive evolution that neural connections had emerged that compelled our animal to believe in a "higher" power, in what we refer to as a god or gods.

As infants in the crib, when we experience pain or fear, we instinctively reach out to our parents for comfort and protection. It seems likely that our cross-cultural belief in a God represents an extension of that same instinct. As Freud expressed this same notion:

> The derivation of religious needs from the infant's helplessness and the longing for the father aroused by it seems to me incontrovertible, especially since the feeling is not simply prolonged from childhood days, but is permanently sustained by fear of the superior power of fate. I cannot think of any need in childhood as strong as the need for a father's protection.[34]

As a result of the selective pressures placed on our species by our awareness of eternal death, neurological connections had emerged that generated an inherent belief in an all-powerful, imaginary father figure whose infinite powers could protect us from death and all that came thereafter. In summary, what I'm suggesting is that at some point in the last two million or so years, during the emergence of the later hominids, a cognitive adaptation emerged that enabled us to cope with our awarenesses of death, while at the same time allowing us to maintain self-conscious awareness. By having this cognitive mechanism selected into us, we were now "wired" to perceive physical death in a much more palatable manner. Once nature had instilled us with neurophysiologically generated cognitive phantoms that could protect us from inevitable death, humans were better equipped to survive their inherent fear of such. According to the religious psychologist Bernard Spilka, "One of the major functions of religious belief is to reduce a person's fear of death."[35] This same notion is further supported by the sentiments of another religious psychologist Mortimor Ostow, "Religion is a natural defense against man's knowledge that he must die."[36]

Sheltered from the perpetual threat of inevitable death, humans could now proceed with the daily routine of maintaining their more

"Earthly" needs. With the emergence of spiritual consciousness, our cognitive functioning had been stabilized to the extent that we could now go on living in a state of relative calm, even amid our awareness of our inevitable demise. This, I contend, is the purpose of a spiritual/religious function. This is its rationale, its reason for being. If all this is true, however, it suggests that God isn't a transcendental force or entity that actually exists "out there," beyond and independent of us, but rather represents the manifestation of an inherited human perception, a coping mechanism that compels us to believe in an illusory reality so as to help us survive our unique awareness of death.

In the remaining chapters, I will provide a variety of arguments as well as the most recent neurophysiological and genetic research that supports such a hypothesis.

CHAPTER 9

THE "SPIRITUAL" EXPERIENCE

"The mystical experience of God has certain characteristics common to all faiths."[37]

—KAREN ARMSTRONG

Having cataloged man's universal spiritual beliefs and practices, there were still several other components to spiritual consciousness I felt needed to be investigated. One such component came neither in the form of a belief nor a practice but rather as a sensation that appears to be cross-culturally experienced by our species.

In his book *Civilization and Its Discontents*, Freud discusses a letter written to him by his Nobel Laureate friend, Romain Rolland. In his letter, Rolland described sensations he experienced that he felt represented "the true source of all religious sentiment." According to Rolland, these sensations:

> consisted in a peculiar feeling, which he finds he himself is never without, which he finds confirmed

> by many others, and which he may suppose is present in millions of people. It is a feeling which he would like to call a sensation of "eternity," a feeling as of something limitless, unbounded–as it were, oceanic.[38]

Romain Rolland was right in presuming he shared this experience with millions. As a matter of fact, archaeological records suggest that every culture from the dawn of man has experienced sensations almost identical to those articulated by Mr. Rolland. Whether it be the born-again experience of the Pentecostal Christian, Hindu samadhi, Sufi fana, or Zen satori, every world culture has described an experience by which individuals claim to feel as if they have been touched by some "higher" truth or power, an experience almost always identified as spiritual, mystical, religious, or transcendental in nature.

Though such sentiments are often evoked within a religious setting, the same experience can be prompted by engaging in such non-religious practices as meditation, yoga, dance, or chant.* Even the contemplation of certain geometric patterns has been reported to evoke comparable experiences. Michael Murphy of the Esalen Institute asserts that the intense focus and concentration elicited in playing sports is itself a form of meditation that can evoke similar sensations. Besides sports, there are many moving meditations such as those used in the Chinese martial art of tai chi; the Japanese martial art aikido; or the fast, rapturous dances of the Sufi mystics. As another example, yoga, with all of its various movements and positions, represents another means by which humans evoke similar sensations. In *Civilization and its Discontents*, Freud describes such an instance:

---

*In 1997, Japanese researchers found that repetitive rhythms have the effect of stimulating our brain's hypothalamus, which evokes feelings of either serenity or arousal in us. This would help to explain part of the underlying mechanics of why dance or chant elicit such "transcendental" feelings in us, demonstrating the connection between spiritual and musical consciousness.

> Through the practices of yoga, by withdrawing from the world, by fixing the attention on bodily functions and by peculiar methods of breathing, one can, in fact, evoke new sensations and coenaesthesias in oneself, which he [Romain Rolland] regards as regressions to primordial states of the mind which have long ago been overlaid. He sees in them a physiological basis, as it were, of much of the wisdom of mysticism.[39]

According to Dan Merkur, author of *Gnosis: An Esoteric Tradition of Mystical Visions and Unions*, mystical experiences fall into various categories. One such category Merkur refers to as theistic mysticism. Here is an experience which "involves feeling the presence of a personified force which intones a 'higher' power. This can take human form (e.g., Jesus), non-human form (e.g., Krishna, Zeus, Ra, Odin, or Yahweh), animal form (e.g., Bear spirit), or a more elemental form (e.g., the wind or Earth spirit)."[40]

Another variant of the mystical experience Merkur identifies as pantheistic mysticism: "Here, one feels that totality of the world is the greatest power and that one can see themselves as part of that totality."[41] Amidst this experience, a person has a sense that he is a part of all that is around him. As described by one of Merkur's subjects, "I felt myself one with the grass, the trees, birds, everything in nature." In his memoirs, Einstein offered a personal account of his own such experience:

> Still there are moments when one feels free from one's own identification with human limitations and inadequacies. At such moments, one imagines that one stands on some small spot of a small planet, gazing in amazement at the cold yet profoundly moving beauty of the eternal, the unfathomable: life and death flow into one, and there is neither evolution nor destiny; only being.[42]

Merkur goes on to list what he considers the five most common symptoms of a mystical experience: "a sense of unity or totality," "a sense of timelessness," "a sense of having encountered ultimate reality," "a sense of sacredness," and "a sense that one can not adequately describe the richness of their experience," a symptom the pioneer of religious psychology, William James, referred to in his *Varieties of the Religious Experience* as the experience's "ineffable" quality.

Other expressions used to describe the mystical experience have included such sentiments as "a feeling of an indissoluble bond, of being one with the external world as a whole,"[43] "a 'higher' experience,"[44] "pure conscious experience,"[45] "cosmic consciousness,"[46] "feelings of unity,"[47] "a greater awareness of a higher power or ultimate reality,"[48] "diminishment or loss of sense of self,"[49] "dissolution of the normal ego; a new kind of ego functioning,"[50] "an altered perception of space and time; ineffable; appreciation of the holistic, integrated nature of the universe and one's unity with it,"[51] and "God-consciousness."[52] Furthermore, such experiences are usually described as evoking feelings of "equanimity; rapture; sublime happiness,"[53] "bliss,"[54] "ecstasy,"[55] "intense positive affect,"[56] "peace and joy,"[57] "a state of assurance,"[58] and "elation."[59]

If one were to look at the whole of these descriptions, they can almost be broken into several categories, each which has been found, as I will go on to show, to correlate to specific regions in the brain. Such sentiments as a "loss of sense of self" or "dissolution of one's normal ego boundaries" describe an experience that is transpersonal in nature, in which personal identity is temporarily suppressed, leaving one feeling detached, egoless, at one with the cosmos.

Another set of experiences involve feelings of "timelessness" and "spacelessness" indicating that normal modes of temporal and spatial consciousness are also suppressed. A third depicts sentiments that are sensual in nature. Terms such as rapture, ecstasy, elation, and bliss all reflect a sensual experience in which normal anxiety dissipates.

As a testament to the impact these experiences have on us, some cultures have created words to describe these sensations. The people of India, for example, have a word, Saccidananda, that appears quite frequently in their sacred and philosophical writings. "This composite Sanskrit word consists of three separate roots: Sat meaning existence or being; Cit, awareness and intellect; and Ananda, bliss."[60]

The fact that so many cultures have described experiencing these particular sensations and in such similar terms suggests that this represents yet another cross-cultural characteristic of our species (i.e., another genetically inherited trait).

> Confirmation of the genuineness of mystical experiences is to be found in the high degree of unanimity observable in the attempts to describe its nature.[61]

Just as all cultures experience sadness, all cultures undergo spiritual experiences. Furthermore, just as the experience of sadness is described in similar terms by every culture, the same is true of spiritual experiences. That all cultures have described sadness in such a similar way indicates that this sentiment is not learned but an inherent part of our human natures. By the same logic, this should hold true of spiritual experiences. And if our capacity to have "spiritual" experiences represents an inherent characteristic of our species, this would imply that such experiences must be generated from some part or parts of our brain, a conviction that is becoming more accepted as new technologies are beginning to offer us actual glimpses into the neuromechanics of human consciousness and, in particular, spiritual consciousness. As expressed by the psychologist James Leuba, "The mystical experience can be explained in physiological terms."[62]

Offering physical evidence to validate this notion, Andrew Newberg and Eugene D'Aquili at the Nuclear Medicine division at the University of Pennsylvania used single positron emission computed tomography (SPECT) scans to observe changes in the neural activity of Buddhist monks. These experiments showed that while the monks were engaged in the act of meditation–in the midst of perceiving themselves as being one with all creation–there was a noticeable change in the neural activity of the frontal and parietal lobes as well as in the brain's amygdala, providing physical confirmation that spiritual experiences can be directly correlated to certain regions of the brain.

When the monks were in the midst of their "heightened" experiences, their brain scans revealed that there was a sudden decrease in blood flow to their brain's amygdala. Being that the amygdala is the part of the brain where fear and anxiety are generated, it makes sense that when the amygdala's blood flow ebbs, our normal fears and anxieties are dissipated, leaving us in a state we cross-culturally describe as being tranquil, euphoric, blissful, serene.

Another part of the brain that the scans showed to be affected by meditation was the parietal lobe, which is where spatial and temporal consciousness is generated. As the blood flow to this area was also suppressed, it makes sense that by meditating, we are left feeling "timeless" and "spaceless."

Lastly, as the frontal lobe has been attributed to generating one's sense of self (Miller, 2001), it becomes clear why a change in blood flow to this region might evoke sentiments often described as a loss of sense of self, or dissolution of one's normal ego. This clearly demonstrates that conscious states, in this case spiritual consciousness, can be reduced to our neurophysiologies. Apparently, we experience such "spiritual/mystical" sentiments not because we are being touched by Heaven or God but because, by focusing our attentions in very particular ways, we can manipulate our neurochemistry, thus altering perception. In support of this notion, *The Comprehensive Textbook of Psychiatry* asserts that "the spiritual contents of consciousness can be accounted for by the effect of excitation of the frontolimbic forebrain."[63]

Providing more evidence to confirm the link between human neurophysiology and religious experiences, Dr. V. S. Ramachandran of UC San Diego's Center for Brain and Cognition Research found that 25 percent of those who suffer from a form of epilepsy that involves activity within their temporal lobes experience a distinct religious fervor moments before they undergo a seizure. Moreover, during their seizures, Ramachandran's patients claimed that "they see God" or feel "a sudden sense of enlightenment."

Dostoyevsky, who suffered this form of epilepsy, offered a description of the experience in his book *The Idiot*: "I have really touched God. He came into me, myself; yes, God exists, I cried. You all, healthy people can't imagine the happiness which we epileptics feel during the second before our attack." In addition, those who suffer from temporal lobe epilepsy have a tendency to be unusually preoccupied with religious concerns, not just during their seizures but also during their everyday lives. In support of this, *The Comprehensive Textbook of Psychiatry* lists "hyperreligiosity" as one of the chief behaviors consistent with temporal lobe epileptics.

To delve deeper into this phenomenon, Ramachandran used skin sensors to compare and contrast individuals' emotional responses to a variety of words and images. Unlike the majority of those tested, who exhibited a heightened sensitivity to sexual language or images, temporal lobe epileptics, who were much less affected by sexual stimuli than the average person, nevertheless showed a heightened, though completely involuntary, response to religious words and icons.

In support of Dr. Ramachandran's findings, Jeffrey Saver and John Rabin of the UCLA Neurologic Research Center found historical documentation to suggest that a significant number of the world's spiritual prophets and leaders were sufferers of temporal lobe epilepsy. The list they composed included, among others, such notable religious figures as Joan of Arc, Mohammed, and the apostle Paul.

Meanwhile, Michael Persinger used a machine called a transcranial magnetic stimulator (a helmet that shoots a concentrated magnetic field at specific areas in the brain) to excite different regions within his own brain. In support of Ramachandran's work, when Dr. Persinger used the device to stimulate his own temporal lobe, he experienced what he described as his first feelings of "being at union with God." When the device was used on volunteer students in a research study, many reported spiritual and mystical experiences, as well as seeing visions of Jesus, angels, and other spiritual deities (meanwhile, some subjects reported near-death experiences as well as alien encounters and abductions,* offering support to the possibility that such perceptions may also relate to temporal lobe sensitivity).

Apparently, the human animal has been "hardwired" in such a way that when we engage in certain acts (e.g., meditation, prayer, chant, yoga, dance, religious ritual or contemplation), it evokes certain perceptions, sensations, and cognitions that we cross-culturally tend to interpret as evidence of some divine, sacred, or transcendental reality. Nevertheless, recent discoveries in the neurosciences contradict such notions by suggesting that religious/spiritual/mystical/transcendental experiences are not manifestations of contact with the divine but rather the manner in which our brain interprets certain genetically derived neurochemical processes.

*Ufology and the belief in extraterrestrial visitations, I believe, constitutes yet another offshoot of the many ways that our spiritual predispostions compel us to believe in some form of a "higher" or alternate power in the universe.

Origins of the Spiritual Experience

If we presume that this sensation we cross-culturally define as either a spiritual, religious, mystical, or transcendental experience represents a genetically inherited characteristic of our species, we must, as always, ask why? Why does our species experience this particular sensation? What is its purpose? How might it enhance our species' survivability? Again, if this series of sensations provided no specific function, it's unlikely it would have emerged in us.

As discussed, spiritual consciousness probably evolved in response to self-conscious awareness, which brought with it, as an unfortunate side-effect, an awareness of death. As a result of mortal consciousness, the human animal would have to live in a state of constant dread unless something could help relieve us of the painful effects of this awareness. If not for the evolution of such a palliative mechanism, it's quite possible our species might not have survived.

One of the ways our spiritual/religious function operates is by generating an inherent belief in supernatural beings, a soul, and the continuity of that soul in what we call an afterlife. As a result of these inherited cognitions, human beings believe they are immortal. In perceiving ourselves as immortal, we are relieved of a great deal of the psychological strain that comes as a result of our unique awareness of inevitable death. But while believing in the existence of a spiritual reality is one thing, experiencing it is totally different.

Though believing in the comforting presence of a spiritual reality, a god, a soul, and an afterlife might help to relieve us of some of our mortal fears and anxieties, humans possess the added benefit of being able to experience euphoric sensations that not only make us feel good (thereby reducing stress levels as discussed in more detail in chapter 12) but also act to bolster our religious beliefs. Because these sensations induced by spiritual/mystical/transcendental experiences are so different from our "normal" modes of consciousness, we tend to interpret them as sublime in nature, as contact with the sacred or divine. The fact that if we close our eyes and focus our concentration on some higher power or god, it alters our neurochemistry in such a way as to transform our conscious experience, and in such an unusual manner, compels us to believe that our beliefs in a spiritual realm are genuine. And with our faith bolstered by these experiences, our mortal fears are all the more diminished.

And so, nature apparently selected a variety of means by which humans can deliberately induce these anxiety-reducing "spiritual" states, which include reciting religious texts, chanting, singing, dancing, meditating, praying, ingesting psychedelic drugs (which will be discussed in greater detail in chapter 10), and in some cases even sex.* Acts such as these have the capacity to trigger a specific series of sensations, ones that we are learning are derived not from any interaction with "divine" forces but rather from electrochemical impulses being generated from within the human brain.

Of the symptoms attributed to this type of experience, one most often described involves a feeling of being one with some greater whole, the dissolution of one's normal ego boundaries. Consequently, in order to understand the nature of this aspect of a spiritual experience, we first need to explore the nature of the human ego, of self-conscious awareness.

---

*It appears that our spiritual centers not only interact with mathematical, linguistic, musical, and moral (chapter 18) consciousness but with sexual consciousness as well. In a variety of cultures, sexual intercourse is viewed as a "sacred" union. Among many primal cultures, the sex act is symbolically reenacted through various fertility rites, often meant to rouse the participants into states of sexual-spiritual ecstasy. In several Eastern cultures, the "sacred" practice of Tantra demonstrates the apparent connection between spiritual and sexual consciousness.

The Ego Function

"The self is a relation which relates itself to its own self."[64]
—Soren Kierkegaard

There is not a healthily functioning human who cannot recognize his or her own reflection. Though most other animals can identify one of their own species, only humans can recognize themselves. Only humans possess a developed sense of self-conscious awareness.*

This unique capacity for self-awareness must represent a trait that emerged sometime during the evolution of the hominids, those creatures that evolved from primates, and of which we are the last surviving species. Since self-awareness represents a cross-cultural characteristic of our species, we can presume that it represents another genetically inherited trait. This would suggest that there exists a group of physiological sites within our brain from which self-awareness is generated. I will refer to this nexus of sites as the "ego function." Furthermore, if such sites exist, it would suggest that there must exist genes that manufacture these parts.

As the "dissolution of normal ego boundaries" stands as one of the primary characteristics of a spiritual experience, not until we understand the underlying physical nature of our "ego" or, more precisely,

*Though only humans possess a capacity for self-awareness, evidence indicates that chimpanzees also possess limited self-perceptual capabilities. In one experiment (Gallup, 1970), chimpanzees were housed in individual cages with a full-length mirror standing outside facing them. For the first few days, the animals screamed at the sight of their own reflections, made threatening gestures, and behaved in a manner consistent with that when chimps are confronted by another of their species. Several days later, however, the chimps' behavior changed. Instead of responding to their reflections aggressively, the caged subjects began to use the mirror as a tool with which to groom themselves, similar to the manner in which humans do, for instance, when we comb our hair. In some cases, the chimps were seen using the mirror to pick food from their teeth. (Monkeys, on the other hand, even after hundreds of hours of being left in the same exact setup, showed no signs of self-recognition.) Once again, given the chimps' evolutionary proximity to our own species, it would make perfect sense that they should demonstrate such incipient self-perceptual capacities.

of self-perception, can we fully understand the underlying nature of spiritual consciousness. As is true of any of our cognitive functions, the ego function is comprised of a group of interactive cognitive parts or processors. Consequently, before we can determine the physical nature of a spiritual experience, we must first understand the physical nature of each of those parts of the brain that pertain to identity and self-awareness. After all, isn't it our identity, our sense of self, that we imagine to constitute our immortal soul?

One of the chief components underlying self-awareness involves something called "episodic" or "autobiographical" memory. Autobiographical memories are those that pertain to one's personal sense of identity, be it one's name, address, family, history, etc. Memories of this sort are believed to be stored in the brain's hippocampus. We believe this because damage to the hippocampus has been cited to precipitate a variety of amnesic states, causing one to forget everything that pertains to self-identity. According to cognitive scientist David Noelle:

> Some amnesics can recall events from early in life, but fail to form new memories for life events. Thus, they may have a coherent sense of self but might feel as if no time has passed since their damage appeared. Other amnesics seem to retain no memory of their past at all. They emotionally report a sense that today is the first day of their lives...that they have just become conscious. Our memories apparently play an important role in constructing a sense of ourselves as unified entities persisting through time. Without these memories, our sense of self seems somewhat disrupted or disturbed.

As V. S. Ramachandran wrote in his book *Phantoms in the Brain*, "If you lost your hippocampus ten years ago, then you will not have any memory of events that occurred after that date."[65]

Two more integral aspects of how we perceive ourselves involve what is referred to as body image and body consciousness. Body image constitutes that part of the human conscious experience by which we perceive our own physical appearance, what we see when we look in the mirror or imagine ourselves. Body consciousness constitutes that part of the human conscious experience through which we perceive our physical presence. For instance, if I raise my arms when my eyes are closed, I have a sensory awareness of my arms being elevated. It has been suggested by Dr. Ramachandran that this particular form of consciousness can be attributed to the right parietal lobe. This deduction is based on the fact that people with damage to their right parietal lobes develop an altered sense of body consciousness. For example, many people with right parietal lesions who are paralyzed on one side of their body often deny their paralysis. They describe such imaginary movements as waving their arm about even though it is clearly immobile. This tendency to imagine illusory or phantom body movements (or to confabulate, as it is referred to by neuroscientists) is a common symptom of those with right parietal lesions. And if our brains can make us sense the presence of phantom limbs, isn't it conceivable that they could potentially compel us to sense the presence of phantom beings?*

Cotard's syndrome, which involves the brain's amygdala, represents another example of a cognitive dysfunction in which the victim suffers from an incapacity to comprehend his own physical being. As a result of damage to one's amygdala, that person may feel alienated or dissociated from his own body or body parts. Someone suffering this syndrome might, for instance, look down at his own arm and suggest that it doesn't feel as if it belongs to him. In more extreme cases,

---

*In regard to this human propensity to sense illusory body movements, there exist distinct similarities between this type of neurophysiologically based syndrome and accounts of what are perceived in a more spiritual context as an out-of-body experience (OBE), otherwise known as a conscious or astral projection (CP or AP). An OBE/CP is most commonly described as a sensation of having one's ego or conscious self leave the physical body and float outward and beyond to another place or, in many cases, to another realm. In light of recent discoveries that reveal that such sensations can be attributed to physical activity taking place within one's right parietal lobe, it's quite possible that it's this same part of the brain–not one's spirit–that is responsible for sensations mistakenly perceived as a CP, AP, or OBE.

a person may even describe feeling detached from his own reflection as he looks upon himself in a mirror. Such dysfunctions as these demonstrate that body consciousness as well as self-conscious awareness are inextricably linked to one's neurophysiology.

More evidence to support an organic explanation of human identity has recently been provided by Dr. Bruce Miller, a neurologist at UCSF who has pinpointed the part of the brain which regulates one's most essential personality components. From one's religious and political views to his likes and dislikes, all originate from a portion of the right frontal lobe (the same region that was shown to receive an altered blood flow during Dr. Newberg's fMRI scans of the meditating monks). This was made evident to Miller when he noticed that people who had suffered damage to this portion of their brain experienced drastic transformations of their core personality, changing everything from their most basic tastes (be it in food, clothing, or music) to their values and beliefs.

Another component of self-identity is contingent on our capacity to make choices. Undeniably, part of how we perceive ourselves is based upon the decisions we make. Should I turn right or left, pick the red or blue one, choose cherry or vanilla? With the advent of modern neuroscience, even our capacity to choose "has been attributed to the limbic system, including parts of the anterior cingulate gyrus. This process connects subjective experience with specific emotions or goals, enabling one to make choices."[66]

Furthermore, "when the amygdala and anterior cingulate gyrus are disconnected, disorders of free will occur."[67] People who suffer this type of cognitive dysfunction become paralyzed in indecision when confronted with options. Such simple tasks as whether to turn right or left at the next corner can render one immobilized. Because these people cannot make spontaneous decisions, their movements appear forced or spastic like that of a malfunctioning automaton.

Based on such findings, it would appear that neither our memory, our rudimentary personality components, nor our capacity for self-conscious awareness are contingent upon the stirrings of some immutable, transcendental component or soul that resides within us but rather on one's neuromechanics. Even so, few people will probably ever really embrace such a reductionistic interpretation of selfhood. This is because just as a planarian turns to the light, humans instinctually believe in free will and the existence of a transcendental soul. And though instincts can be suppressed, they can never be extinguished.

In the coming pages, as I speak of an ego function, it is not to be confused with either Freud's or Jung's definitions of this same term. Though I agree with Jung that the ego represents that part from which our sense of self is generated, he viewed consciousness as a manifestation of the ambiguous "mind," whereas I view it as a purely physical phenomenon, the product of electrochemical signals being transmitted throughout the brain. In a sense, I am seeking to biologize Jung's conception of ego-consciousness.

But before we take such a mechanism as an ego function for granted, we must first ask: If our capacity for self-awareness is physiological in origin, what is its purpose? How might such a function increase our species' survivability? Again, if no such purpose can be determined, it's not possible to justify its theoretical existence.

As infants, we do not yet possess a developed sense of self. At this early stage in development, a human being cannot distinguish its own existence from the world around it. As Freud expressed it, "an infant at the breast cannot distinguish his ego from the external world as the source of sensations flowing in upon him."[68] What this means is that when we are born, the ego function, like our language function, for instance, has yet to be developed and exists in a latent stage. Self-conscious awareness is therefore something that emerges in us sometime after we are born.

Based on experiments he performed, the developmental psychologist Jean Piaget came to a similar conclusion–that humans are born without any recognizable sense of self. Studying the cognitive development of children, Piaget observed that before the age of two, children possess little, if any, sense of self-conscious awareness. Piaget classified this pre-self-aware phase of our existence as the "sensorimotor stage" of human development.

According to Piaget, it is between the ages of two and seven, during what he referred to as a person's "preoperational stage," that a child learns to recognize his own image as well as to develop a sense of his own self as an autonomous being, separate and unique from his mother and the rest of the world. As the child becomes conscious of his autonomy, he develops a sense of self-responsibility. He realizes that he must learn to fend for himself. It is during this stage that a child also learns to feed himself, wash himself, and go to the bathroom by himself. And so, slowly but surely, we grow from utterly dependent to independent (or, at least, interdependent) beings.

As a child's sense of self unfolds, he develops an instinct for self-preservation, a desire to sustain and protect his newfound self. The stronger his sense of self, the more he will want to care for himself. As a result of our capacity to recognize ourselves, we have become the only species in which an individual can develop a genuine bond with himself. We are consequently nature's first narcissistic creatures, the first animals to possess a capacity for self-love. In a sense, one could say that we have the capacity to develop the equivalence of maternalistic feelings for ourselves. And so, with the same fervor and intensity with which a mother will love, care for, and defend her young, human beings can love, care for, and defend themselves. This, I believe, constitutes one of the chief advantages of self-awareness.

It is for this reason that the preoperational stage plays such a critical role in our emotional developments.* The conditions under which a

---

*It is during the preoperational stage in our natural cognitive developments that spiritual consciousness first emerges in us (Elkind, 1961; Decochny, 1965; Long, Elkind, Spilka, 1967). Similar to the manner in which we are born without linguistic, moral, or mathematical consciousness, humans are born without any sense of spiritual or religious consciousness. It is during this stage, however, that humans have their first conceptions of gods, spirits, souls, and afterlives. It is also during this same stage that we first develop self-conscious awareness as well as an awareness of our own mortalities, which may play a role in the emergence of our spiritual sensibilities. In support of this notion, Dr. K. Tamminen, in his book *Religious Development in Childhood and Youth*, reported that feelings of closeness to God among seven- to eleven-year-olds is generally linked to "situations of loneliness, fear, and emergencies–such as escaping or avoiding danger–or when they were ill."

child is raised at this time (what is often referred to as one's formative years) will determine the manner in which one will learn to perceive himself. If a child is raised in a nurturing and loving environment, he will develop a positive self-image, in which case he will learn to love and cherish himself. The more a human loves and cherishes himself, the more effectively he will fend for himself. If, on the other hand, a child is raised in an unhealthy environment, he will likely develop a negative self-image, which may eventually foster a host of self-destructive tendencies. We call such unhealthy tendencies neuroses. Neuroses are therefore the behavioral consequences of an unhealthily developed self-image or ego function.

Another benefit of self-conscious awareness is that it grants us the ability to modify ourselves. Because we can perceive ourselves, we can recognize our own shortcomings. This affords us the capacity to turn our weaknesses into strengths. For example, though humans aren't born with the capacity to fly, should we perceive this as a shortcoming, we can build ourselves flying machines. Though we might not be born the fastest creatures on Earth, by recognizing this as a shortcoming, we can build ourselves racing machines. Should another ice age strike, we won't need to wait millions of years for nature to select thicker coats of hair for us but can sew ourselves one within a few hours' time. Pertaining to self-image and body consciousness, should a person feel, for instance, that he is dangerously overweight, he can diet. In this way, our species, and ours alone, has the capacity to modify itself, to compensate for any physical deficit and, consequently, to transform it into a potential strength, thus rendering us the most versatile and resilient of all Earth's creatures.

So how does the ego function work? The ego function acts as the body's control center (what neuroscientists refer to as our executive processor). If the body were a ship, the ego would be its captain. If the body is our temple, the ego is our high priest. Whereas the heart is responsible for the pumping of blood, the ego is responsible for the supervision of our body's entire upkeep. It does this by acting as our body's personal manager, that part of us which is responsible for making all decisions. Should I seek food first or shelter? Shall I turn right or left at the next corner? All such decisions are made not by our kidneys, livers, or even language centers within the same organ, but by those parts from which our sense of self as well as our capacity to make decisions–our executive processor–is generated.

As stated, the ego function is responsible for the body's entire upkeep. For example, when we feel hunger, it is our ego mechanism that informs us that we must provide food for ourselves. As the manager of our existences, it is consequently the ego that must bear the brunt of all of our physical needs and responsibilities. When hunger must be assuaged, it is not the heart's, or the stomach's, or the kidney's responsibility, but the ego's to find the body its next meal.

When an individual feels pain, it is his or her ego that suffers. For example, if someone were to stick a pin in my hand, it is not my hand that endures the pain, per se, but "me," my ego (brain) that registers the experience. Remove or suppress a man's ego mechanism and you can turn him into a human pincushion and he won't feel a thing (as in the case of someone in a coma who, though their pain receptors are in perfect working condition, because they are "brain dead," are immune to any such pain). My hand doesn't experience pain; I do. It is not my tongue that tastes the apple but me, my ego, that does.

Consequently, the ego is not only the seat of self-perception, but it is that organ responsible for all decision-making, and therefore for essentially most everything we do. Should I need to procure a meal or find shelter, it is I, my ego, that bears the brunt of this and every personal responsibility, every choice I need to make. It is therefore my ego mechanism that must bear the brunt of all my consequent anxieties, including that most debilitating anxiety of all which comes as a result of our species' unique awareness of death.

As is true for any organ, when physical strain becomes too great, that organ becomes susceptible to mechanical breakdown. If I lift too much weight, I may tear a ligament. If I overexert my heart, I may suffer a heart attack. For every body part we possess, there exists a threshold of strain before it will break. Consequently, if ego-consciousness is based in some specific neurophysiological mechanism, then as is true of any part of us, if overextended, it can and will break. Consequently, if our ego mechanism didn't possess some means by which to relieve itself of the excess strain that comes with our awareness of death, it would be at risk of suffering a physiological breakdown. And when the ego breaks, all is lost. After all, what good is a ship once it has lost its captain?

What therefore happens when our ego must bear the overwhelming strain that results from our species' unique awareness of death? Imagine having to experience one's entire life under the same conditions that a rabbit experiences when cornered by a mountain lion–its body pumped with adrenaline, its heart palpitating, its muscles tensed, its brain surging with painful anxiety. Imagine having to experience this same anxious strain all day, every day, for the rest of one's life. Under such stressful conditions, how could one survive? How would one be able to perform any of life's normal, daily functions? It would be impossible (if in doubt, ask someone who suffers a severe anxiety disorder). Pitted against the terminal threat of imminent death, we are left in a perpetual state of existential paralysis, unable to either fight or escape the object of our fear.

Imagine the burden such a condition must have placed on our newly emergent ego mechanisms, exactly the type of undue strain that would render any physiological function susceptible to breakdown. If our egos were to continue to function under such conditions, some cognitive mechanism had to be selected in us that could relieve us of at least some of this excess strain. Had nature not provided us with such a device, it's possible that our species might have suffered a cognitive meltdown that might have rendered us extinct.

It was at this point in our species' evolution, during the emergence of our self-perceptual capacities, that the forces of natural selection provided us with a mechanism by which our ego functions could endure the overwhelming strain that came as a result of our debilitating awareness of death. I refer to this mechanism as the "transcendental" function.

The Transcendental Function

"The peculiar structure of the human ego results from its incapacity to accept reality, specifically the supreme reality of death."[69]
–NORMAN O. BROWN

"Sometimes as I drift idly on Walden Pond, I cease to live and begin to be."
–HENRY DAVID THOREAU

In order to save our ego function from the severe strain caused by our constant awareness of death, nature could have done one of several things. As one solution, it could have displaced the strain onto some other part or organ, something that would only have proved to be equally damaging (this tends to happen to a certain extent anyway, as psychological stress has been cited to play a key role in the development of a number of ailments and illnesses). As mentioned earlier, "nature" could have weeded out the more intelligent of our species, therefore eradicating our capacity for self-conscious awareness and with it our awareness of death. Compromising our intelligence, however, would most likely have been even more damaging.

Another strategy "nature" could have employed would have been to select a mechanism that would enable us to temporarily suppress our ego function as a means to dispel the debilitating anxiety incurred by life's daily stresses as well as the more severe strain caused by our awareness of death. By providing us with such a mechanism, the human animal would be less susceptible to suffering a biopsychological breakdown.

If we recall the descriptions of a spiritual/mystical experience, there was an entire set that suggested a suppression of the ego function. Such expressions as "a loss of sense of self" or "dissolution of one's normal ego boundaries" reflect states in which the ego function is held in abeyance. With the ego temporarily shut down, there is no longer a coherent "I" through which to experience pain or anxiety. Instead, we are left feeling egoless, detached from any coherent sense of self, a state universally depicted as cosmic, boundless.

With this capacity to disengage our ego function, we are given a temporary reprieve from the excess strains of daily existence. During this experience, we retreat to an altered state similar to that into which we were born, one in which we can no longer differentiate between our own internal reality and the external world around us. As Freud expressed this same idea:

> Our present ego feeling is, therefore, only a shrunken residue of a much more inclusive–indeed, an all-embracing–feeling which corresponded to a more intimate bond between the ego and the world around it.
>
> If we may assume that there are many people in whose mental life this primary ego-feeling has persisted to a greater or lesser degree, it would exist in them side by side with the narrower and more sharply demarcated ego-feeling of maturity, like a kind of counterpart to it. In that case, the ideational contents appropriate to it would be precisely those of limitlessness and of a bond with the universe–the same ideas with which my friend elucidated the "oceanic" feeling.[70]

With our ego function temporarily suppressed, we experience feelings of "being one with the external world as a whole," of "cosmic" or "God" consciousness. And so, with our captain momentarily relieved of its duties, all anxiety is temporarily abated. In such a state, we feel freed from all sense of personal responsibility, disassociated from normal concerns, fears, and anxieties, immune to physical pain and suffering, which is why the spiritual experience is often described as generating feelings of "euphoria," "rapture," "bliss," or "tranquility." Being that we, as a species, are predisposed to believing in the sacred and the sublime, we tend to interpret such sensations as evidence of God, or at least of some transcendental reality.

In a wakened state, humans experience a brain wave frequency of about thirteen cycles per second, what is referred to as a Beta wave. When we close our eyes and focus our attentions inward–when we meditate–our brains shift to an Alpha state of eight to twelve cycles per second. In addition, it has been shown that when a person is in the midst of an Alpha brain wave state, there is a tendency to be less responsive to physical pain.

It is a common claim of individuals in the midst of a meditative or trancelike experience to be impervious, or at least less susceptible, to pain. Whether demonstrated by someone lying on a bed of nails or walking across hot coals, the evocation of a meditative or mystical experience seems to make us at least partially immune to physical pain. According to studies done on Yogis, those practicing meditation "claim to reach a state [known as mahanand (ecstasy)] that surpasses the experience of pain."[71]

And how is it possible that we can immunize ourselves from physical pain? It's because when we suppress our ego function, there is no conscious self through which to experience physical pain or anxiety. With the ego, our cognitive captain, physiologically suppressed, there is no "I" through which to experience such negative sensations.

Apparently, the act of meditation has physiological consequences. As a matter of fact, the last few decades of research have yielded so much evidence linking mental processes to autonomic, immune, and nervous system functioning that it's spurred the creation of a whole new discipline known as "psychoneuroimmunology."

In 1968, Harvard cardiologist Herbert Benson was contacted by practitioners of transcendental meditation (TM) who asked him to test their ability to lower their own blood pressures. Not only did Benson find this to be the case, but subsequent studies have reported that the use of TM is associated with increased longevity and reduction of chronic pain (Kabat-Zinn et al., 1986); reduction of high blood pressure (Cooper and Aygen, 1978); reduced anxiety and reduction of serum cholesterol level (Cooper and Aygen, 1978); reduction of substance abuse (Sharma et al., 1991); treatment of post-traumatic stress syndrome in Vietnam veterans (Brooks and Scarano, 1985); blood pressure reduction in African Americans (Schneider et al., 1992); and lowered blood cortisol levels initially brought on by stress (MacLean et al., 1992).

Based on the above research, it appears that not only can the act of meditation help to immunize us from pain, but it can also decrease life-threatening anxiety levels, which, in turn, reduces our chances of incurring certain physical ailments.

Another of the primary symptoms attributed to a spiritual/transcendental/mystical experience involves feelings of timelessness and spacelessness. Once again, Newberg's neural SPECT scans revealed that the act of meditation causes a decrease in blood flow to the brain's parietal lobe. As the parietal lobe is that part of the brain responsible for orienting us in time and space, by having this part of us relaxed, we experience a feeling of timelessness, spacelessness, disassociated from our normal perspective of reality. Add to this the fact that our frontal lobe becomes excited during meditation. As the frontal lobe regulates focus and attention, the spiritual experience feels even more intensified. As stated by Eugene D'Aquili in his book *The Mystic Mind*, "This results in the subject's attainment of a state of rapturous transcendence and absolute wholeness that conveys such overwhelming power and strength that the subject has the sense of experiencing absolute reality. This is the state of absolute unitary being. Indeed, so ineffable is this state that for those who experience it, even the memory of it carries a sense of greater reality than the reality of our everyday world."

Because we are "wired" to ascribe spiritio-religious significance not only to the world around us but also to our own experiences, we are predisposed to interpreting this type of altered state of consciousness as divine or transcendental in nature, what Otto Rank referred to as a "numinous" experience. Nevertheless, regardless of how we may be disposed to interpreting such states, modern neural-imaging technologies, which have allowed us to glimpse into the biological nature of cognition, have revealed that what we perceive as spiritual/mystical/transcendental experiences can be reduced to the workings of our basic neurobiology–this and nothing more. Though we have no evidence whatsoever of the existence of any spiritual reality, there is real, hard evidence to suggest that spiritual experiences are strictly physical in nature, the product of human cognition. Apparently, it is not a transcendental soul that we possess but rather a physical brain. As the neurobiologist Steven Rose expressed this same notion:

> It is highly probable that in due course it will be possible to explain the "mystic experience" in terms of neurobiology; it is highly improbable that neurobiology will ever be explained in terms of "the mystic experience."[72]

Chapter 10

Drug-Induced God

"Psychedelic drugs have been used to stimulate religious experience since the dawn of history."[73]

—C. D. Batson

"Religion is the opiate of the masses."

—Karl Marx

Besides engaging in such practices as prayer, chant, dance, yoga, or meditation, many world cultures have used psychedelic drugs as yet another means through which to evoke a mystical experience. In the words of the cultural anthropologist Robert Jesses:

> The use of psychedelic sacraments in shamanic and religious practices is found throughout history. The word entheogen, used to describe certain plants and chemicals used for spiritual purposes, emphasizes this long-established relationship.[74]

The sacred drink of soma used by the Vedic Hindus, the morning glory seeds and mescaline ingested by Native Americans, the sacred mints of the Greek mystery religions, the use of cannabis by the Scythians, the yaje or ayahuasca of the Amazonian jungle peoples, and the iboga of the peoples of equatorial Africa are all examples of psychedelic drugs used to evoke a spiritual experience. Because of the universal nature of this phenomenon, the word entheogens–meaning "God generated from within"–has been created to describe this class of "God-inducing" drugs. To the ancient Aztecs, the connection between entheogens and the spiritual realm was so clear that they referred to peyote as the "divine messenger" and psilocybin as "God's flesh."

It is so widely recognized that certain drugs can stimulate a spiritual experience that some secular governments, which normally forbid the use of drugs, have legalized certain entheogens when ingested as a religious sacrament. "In 1994, the U.S. government enacted the American Indian Religious Freedom Act Amendments, providing consistent protection across all fifty states for the traditional ceremonial use of peyote by American Indians...In its report on the 1994 legislation, a U.S. House of Representative's committee reported that 'peyote is not injurious,' and that the spiritual and social support provided by the Native American Church (NAC) has been effective in combating the tragic effects of alcoholism among the Native American population."[75]

From William James's experiments with nitrous oxide to Aldous Huxley's experiments with lysergic acid (LSD), it is widely noted that certain plants and/or chemicals can induce experiences indistinguishable from certain mystical states. Stanislov Grof, in his work *Realms of the Human Unconscious: Observations from LSD Research*, cataloged the experiences of individuals who were administered experimental doses of LSD. Based on his studies, Grof found that the symptoms described by those who had taken the drug were nearly identical to those who had undergone a mystical experience.

But how is it that a drug could have the ability to rouse such feelings as these in us? How is it possible that chemicals can have the capacity to induce sensations as allegedly sacred and sublime as a spiritual or transcendental experience? What does this say about such drugs? Or, more significantly, what does this say about a spiritual/transcendental experience?

In order to answer such questions, we need take a look at the drugs themselves. As we know, all drugs, including the psychedelics, or entheogens as they are now called, are always the same in regard to their molecular structure.* This is true of any drug. For example, on a molecular level, aspirin is always aspirin; penicillin is always penicillin. Accordingly, the same rule must also apply to each of the various entheogenic drugs. In other words, the chemical makeup of any entheogenic drug represents a constant. The atomic structure of an LSD molecule is the same whether ingested in Bangkok or Bolivia, at sea level or on top of the Himalayas.

The same can be said, more or less, about human physiology. Granted, though there is a certain degree of variance among individuals within our species, underlying this diversity is a distinct physiological uniformity. Since we are dealing with two constants–same drug, same physiology–it's no surprise that entheogenic drugs should have this same particular effect on individuals from such a diverse range of cultures. This still leaves us with the crux of the problem, which is: why do these drugs have this particular effect on us? Why do they have a distinct tendency to elicit what we refer to as spiritual/mystical/transcendental/religious experiences?

No drug can elicit a response to which we are not physiologically predisposed. Drugs can only enhance or suppress those capacities we already possess. They cannot create new ones. For example, the fact that we possess the capacity for sight–that we possess the physical

*In regard to the molecular composition of many of the entheogenic drugs, it is no coincidence that, in many cases, they are nearly identical in structure to certain neurotransmitters–those chemicals that play an integral role in the chemical transmission of impulses between neurons (nerve cells). For instance, whereas the entheogenic drug mescaline is almost identical in its molecular composition to the neurotransmitter noradrenaline, a molecule of psilocybin, more commonly known as "magic mushrooms," is almost identical in composition to a molecule of the neurotransmitter serotonin.

mechanics to "see"–means that it is within the realm of possibility that a drug would be able to either enhance or suppress one's visual capacities. The fact, however, that we do not possess the physical capacity to fly, for instance, means that no drug can ever enhance or suppress our nonexistent powers of flight. Again, a drug can only affect us as much as we possess some physiological mechanism that might be receptive to a drug's particular chemistry.

The fact, for instance, that novocaine has the universal effect of desensitizing one to pain means that we must possess pain receptors that are capable of being suppressed. In the same way, the fact that psychedelic drugs have a cross-cultural tendency to stimulate experiences we define as being either spiritual, religious, mystical, or transcendental means we must possess some physiological mechanism whose function is to generate this particular type of conscious experience. If we didn't possess such a physical mechanism, there's no way these drugs could possibly stimulate such experiences in us. In essence, the fact that there exists a certain class of drugs–molecular combinations–that can evoke a spiritual experience supports the notion that spiritual consciousness must be physiological in nature. Herein lies the basis for an ethnobotanical argument against the existence of either a spiritual reality or a soul.

CHAPTER 11

THE "SPIRITUAL" GENE

> "The idea of men's receiving an intimation of their condition with the world around them through an immediate feeling sounds so strange that one is justified in attempting to discover a genetic explanation of such a feeling."[76]
>
> —FREUD

Almost as old as the psychological sciences themselves, the debate between nature versus nurture endures: Is human behavior learned or innate? While strict behaviorists see our environments as the determining factor underlying all human action, behavioral geneticists look for the influence that our genes have over the same. Though there is little question that the human animal is shaped by a combination of both of these two interactive forces, the more we learn about genetics and neurophysiology, the more we are discovering exactly how much our genes really do influence our perceptions, cognitions, behaviors, and emotions.

Of the approximately 100,000 genes* that account for the human body, it is surmised that "50,000 to 70,000 are involved in brain function."[77] Such numbers attest to the pivotal role that the human genome plays in our neurophysiological makeups. Moreover, "at birth, a baby's brain contains 100 billion neurons, roughly as many nerve cells as there are stars in the Milky Way."[78] Of these approximately 100 billion neurons with which we are born, "there already exist more than 50 trillion connections (synapses)."[79] What all this amounts to is that before we even have a chance to be influenced by our environments, over 50 trillion connections have already been established within our brain, connections that will inevitably play an essential role in our psychological, emotional, behavioral, and intellectual developments. As a matter of fact, our genes play such a pivotal influence in human behavior that "scientists now estimate that genes determine about 50 percent of a child's personality."[80] Though experience may represent the chief architect of human behavior, it seems our genes constitute its foundation.

With all this in mind, isn't it therefore possible that our genes might play a determining role in an individual's spiritual and/or religious development? According to recent genetic studies, they do play a role–and a significant one at that.

Two of the most effective methods used by science in its search for clues through which to determine the influence of genes on behavior is through the use of twin and adoption studies. In adoption studies, scientists observe the behavioral differences and similarities between genetically related individuals who are raised apart. Even more effective, however, is to compare the results of adoption studies between fraternal (dizygotic or DZ) versus identical (monozygotic or MZ) twins.

Imagine, for instance, that among a thousand sets of fraternal (DZ) twins separated at birth and raised apart, fifty sets grow to have similar musical abilities and tastes. Now imagine that among a thousand identical (MZ) twins separated at birth and raised apart, four hundred sets grow to have similar musical abilities and tastes. In such a case, this

*More recent estimates propose that the human genome is comprised of a much smaller number than had previously been speculated and is closer to approximately 34,000 genes as opposed to 100,000. Nevertheless, we can still presume that at least half of our genes are dedicated to the creation of our neurophysiological makeups.

would suggest that genes probably play an essential role in determining musical ability and taste. "Generally, a greater MZ than DZ similarity for a particular trait is considered evidence for a genetic contribution to the etiology of that trait."[81]

In comparing and contrasting a wide array of specific religious behaviors of "eighty-four identical and non-identical twins reared apart and 821 twins reared together, Waller and his colleagues (1990) came to the conclusion that religious attitudes and interests are genetically influenced."[82]

In another study conducted at Virginia Commonwealth University on thirty thousand sets of twins–the most ambitious twin study to date–researchers concluded that "although the transmission of religiousness has been assumed to be purely cultural, studies have demonstrated that genetic factors play a role in the individual differences in some religious traits."[83] This same team of researchers went on to say that whereas "religious affiliation is primarily a culturally transmitted phenomenon, religious attitudes and practices are moderately influenced by genetic factors."[84]

Yet another study (Kendler, Gardner, & Prescott; 1997) offered that "A twin study of females reports that genetic factors influence personal devotion–a factor including importance of religious beliefs, the frequency of seeking spiritual comfort in difficult times, and the frequency of prayer."[85]

The University of Minnesota, which conducted its own twin study, concluded that "Studies of twins raised apart suggest that 50 percent of the extent of our religious interests and attitudes are determined by our genes."[86]

Based on the cumulative results of twin studies such as those mentioned, it appears that our genes have an undeniable influence on religious behavior.* Herein lies the basis for a genetic argument against the existence of a spiritual reality and for the existence of a spiritual/religious function...a "God" part of the brain.

*As scientists continue to unravel and decipher the contents of the human genome, perhaps there will come a time when we will have knowledge of precisely which genes are responsible for those parts of the brain that give rise to religiosity and spiritual consciousness. In order to accommodate this new field, the sciences may have to look toward a whole new discipline–a new geno-theology–for its answers.

CHAPTER 12

THE PRAYER FUNCTION

> "People who reported increased spirituality described the presence of an energy, a force or power–a god–that was beyond themselves. It was the people who felt this presence who rated the greatest medical benefits regardless of their faiths."[87]
>
> –HERBERT BENSON

Every religion encourages the act of prayer on a daily basis. For Muslims, it is required that all men pray five times a day. In hospitals, temples, sports arenas, and cemeteries around the globe, all cultures beseech some supernatural entity for aid and assistance in dealing with life's hardship and suffering. Few people have never prayed, whether out loud or silently to themselves. As a matter of fact, the act of praying is so universally apparent, it would be hard not to call it an instinct.

Moreover, it is widely recognized that the act of prayer possesses distinct healing properties. This notion has been so well-documented that there are entire shelves in today's book stores devoted to this particular topic, usually under the heading of what are dubiously referred to as the "new age" sciences. Though many of these books offer some spiritual/mystical explanation to this phenomenon, I, on the other hand, will only address the subject insofar as I can offer a physiological interpretation of this phenomenon.

From every corner of the globe a variety of cultures have spoken or written of the healing properties of prayer. According to Herbert Benson, "Many cultures have named and believed in a mysterious healing energy. The ancient Egyptians called it 'ka,' the Hawaiians, 'mana,' the Indians, 'prana.'"[88]

That this phenomenon seems to be cross-cultural in nature suggests that we are dealing with yet another genetically inherited characteristic of our species. Consequently, it follows that there must exist some physiological mechanism or series of mechanisms responsible for enabling this healing capacity we apparently possess.

Evidence indicates that praying expedites the time it takes to recover from sickness or surgery. Apparently, through the act of prayer, humankind possesses a capacity to heal wounds, cure illness, and prevent disease. But how might such a mechanism work in us? How is it possible that the act of prayer possesses such unusual healing properties? Is this the work of miracles or is it simply another genetically inherited physiological response to a specific stimulus?

As we know, the human body is an interactive network of organs. If one organ is not functioning properly, the rest of the body suffers. It is the job of the kidneys, for instance, to filter toxins from the body. If the kidneys are not functioning properly, these now unfiltered toxins will have an adverse effect upon the rest of the body. As another example, when a person has a healthy heart and circulatory system, the body's tissues and organs are provided with an ample supply of oxygen, allowing each of those parts to operate at maximum capacity. An inefficient heart or circulatory system will therefore most likely have some adverse effect on every other part us.

As this principle applies to all of our organs, it also applies to our brain. Therefore, if the brain, which is the body's control center, is not functioning at maximum capacity, neither will the rest of the body. Of the many functions for which the brain is responsible, one is to channel anxiety. If the brain is not performing this function properly, this, too, will have an adverse affect on the rest of the body.

In its healthiest form anxiety works to one's advantage as it is meant to heighten one's responsiveness to urgent situations. In its unhealthiest form, however, poorly displaced anxiety has been found to precipitate panic attacks, nausea, sleeplessness, diarrhea, hair loss, ulcers, palpitations, muscle tension, premature symptoms of aging, migraines, loss of appetite, a variety of eating and addictive disorders, depression, schizophrenia, mood disorders, and susceptibility to colds, flu, virus, and even cancer, to name just a few examples. All of these things place even more strain on the body, which subsequently only acts to evoke even more anxiety in us.

As noted, it is the function of the ego mechanism to oversee the upkeep of our entire body. It is therefore also the ego mechanism that must carry the burden of all those anxieties that come with this vast responsibility. Consequently, there is a great deal of strain placed on this part of our physiologies. In cases in which the ego mechanism cannot adequately displace such strains, it will not be able to perform at maximum capacity.

In keeping with the principle that the body works as an interactive network of organs, if the ego mechanism is not functioning at maximum capacity, neither can the rest of the body. Any strain that the ego mechanism cannot properly handle may end up being displaced onto some other part of us. As most of us possess some part of our body that is more vulnerable than the rest, it is often this part that will suffer the effects of our excess anxieties. This same principle also applies to those organs that are vulnerable because they might either be ailing or in the process of recovering from some wound, disease, or surgery. As such parts are either indisposed or in the process of healing, they are most vulnerable to the adverse effects of our excess anxieties.

Consequently, by reducing anxiety levels, this would expedite the recovery process. In addition, if anxiety can weaken any part of us, it can also weaken our immune systems. By reducing anxiety levels, we therefore optimize our immune systems, thus expediting the healing process even more.

Apparently, by focusing our attentions on what we perceive as the transcendent–that is, by praying (or meditating)–our species has the capacity to alter our physiologies in such a way that we can reduce stress levels, prompting a chain of healing responses upon the body.

But what exactly is the mechanism by which prayer reduces stress? It seems that the human animal has an inherent propensity to believe in supernatural beings who have powers that far surpass our own. In times of adversity, humans have a tendency to turn to these "higher" powers for aid or assistance. Because we believe that these same gods have created all that exists, we believe that, as our creators, they possess certain maternalistic/paternalistic feelings for us. For this reason, we believe that when we solicit our gods for help–when we pray–those same gods will come to our assistance. Just as our parents were there to care for and protect us as children, we instinctively believe that our gods are there to care for and to protect us as adults. As noted earlier, it may very well have been a natural extension of our instinct to seek parental protection that spurred the neurophysiological emergence of a belief in a god or gods.

Because we instinctively believe in the existence of such supernatural forces who possess both the power as well as the inclination to assist us, we are compelled to pray to these forces. Because we instinctually believe that our solicitations will be answered, our anxiety levels are diminished, thus relieving us–our ego functions–of some of our excess psychobiological strain. By diminishing our anxieties–by relieving our ego functions of undue strain–the rest of our bodies, including our immune system with all of its regenerative powers, can function at maximum capacity. With this accomplished, not only will there be less strain displaced on an ailing or recovering organ, but with our immune systems able to operate at a greater capacity, they, too, can more effectively help to complement the healing process. This, I believe, represents the underlying reason that the act of prayer engenders the healing properties it does.

Though there is no definitive proof to support such an assertion, I would go as far as to surmise that children raised by neglectful parents are generally not as healthy as their properly nurtured counterparts. When we are raised in a loving environment in which we are made to feel secure, we are that much less plagued by fear and anxiety. Moreover, those who were neglected as children, I would imagine, are more likely to grow to be more prone to disease and sickness as adults. Just as children with little parental support are more prone to sickness, adults who have no spiritual support–no gods to turn to for hope or assistance–are probably more prone to illness as well.

Though this may help to explain the physical origins underlying the healing and regenerative properties inherent in the act of prayer, how might this account for more radical instances of faith healing in which people have been instantaneously cured from such physical handicaps and disabilities as blindness and paralysis? Though the vast majority of faith healings have been shown to be spurious, at the same time, I believe it's conceivable that, in rare cases, a person could be instantaneously cured of a serious illness or handicap.

As the body's control center, the brain plays an influential role in nearly every bodily function. Since a strained ego mechanism can interfere with the workings of every other part of our bodies, it can also adversely affect all of our other brain functions. Moreover, as the entire nervous system converges at the brain, a distressed brain can affect every single part of the body. When we walk, for instance, it is not because our legs have decided to move of their own accord but because we, by virtue of our brains, have directed them to do so. Consequently, a distressed brain can, theoretically, render a person with perfectly functional legs paralyzed. This is the nature of any psychosomatic illness, one that originates not in the afflicted body part itself but from within the workings of the central nervous system, the brain. Consequently, it is not the afflicted body part that needs attention but the central nervous system, which, as discussed, is highly susceptible to the influences of psychological strain (anxiety). It is for this reason that people who suffer from psychosomatic illnesses are usually cured not with the aid of conventional treatment or medication but rather by being relieved of those excess anxieties that have interfered with the normal operations of some body part.

This is why placebos often possess the healing properties they do. By simply believing that another person has the capacity to cure one's imaginary illness, this can reduce that person's anxiety levels to the extent that it will allow that person's ego function to operate more effectively, thus enabling the rest of that person's nervous system to do the same. Consequently, someone who is psychosomatically paralyzed might regain the use of his legs merely by reducing his anxiety levels based simply on the power of his faith.

This type of response is most dramatically realized when facilitated by the techniques of those referred to as faith healers. Take, for instance, the example of the blind man at the revival meeting who has his vision suddenly and "miraculously" restored by the work of a faith healer. In the few documented cases of those to whom this has actually occurred, according to the "afflicted," no medical doctor was ever able to find an organic cause for his condition and therefore no medical cure could provide relief for his disorder. This is because a psychosomatic illness doesn't originate from the allegedly "sick" body part but instead stems from a debilitated ego function.

Unresolved childhood trauma or repressed guilt often constitute the cause of such strain. Consequently, past memories can carry potent and turbulent emotional content which can disrupt the workings of one's ego function. In order to protect our "executive processor," something without which we would be completely incapacitated, the body reacts by displacing such strains onto some other part of us, perhaps even as a means to distract us from our psychological pain. And so, in order to avoid some painful memory that would cause us to completely break down, we instead only partially break down by becoming blind, deaf, mute, or crippled, each disorder a testament to the power of memory.

This brings us to our faith healer who, though he is incapable of performing miracles, is very adept at tapping into one's prayer function, thus allowing such psychosomatically ill individuals to vent their excess anxieties through the evocation of prayer. By invoking the prayer function, the faith healer is really just facilitating a cerebral catharsis in someone who is stricken with excess anxiety. In this way, the faith healer works like a placebo. By helping to rouse a psychosomatically ill person's built-in capacity to have faith in a god that can "heal" and "save," the suffering person is relieved of a great deal of his anxieties. Once this is accomplished, the strain that was previously displaced on the person's nervous system is relieved to the extent that the psychosomatically ill person may find himself suddenly and miraculously "healed" of whatever it was that ailed him. In essence, a faith healer helps to relieve a disabled individual of those excess anxieties that have precipitated some psychosomatic illness.

What I mean to demonstrate by all of this is that when we are cured or healed through acts of prayer, it is not the result of miracles but rather the consequence of a purely physiological response to having one's anxiety levels diminished. The fact that all cultures have spoken of the healing properties of prayer leads me to believe that our species possesses a distinct set of prayer-responsive mechanisms that exist within our brains all meant to enable us to endure the psychological strain caused by the hardship of life and the certainty of death.

CHAPTER 13

RELIGIOUS CONVERSION

"To whom is the Lord revealed? He is despised and rejected of men, a man of sorrows and acquainted with grief."

—OLD TESTAMENT, BOOK OF ISAIAH

When we speak of a person being "born again," we are generally referring to someone who has undergone a religious conversion. When we see someone who has spent his entire life leading a secular existence suddenly devote himself to an organized cult or religion, this is usually the result of a religious conversion. When someone with whom we used to go to bars and baseball games is suddenly spending his or her days handing out religious pamphlets in the streets, shouting to every passerby that "Jesus saves!" or "Krishna is enlightenment!" this, too, is most likely the result of a religious conversion.

Perhaps we've known someone close to us who has undergone such an abrupt personal transformation, or perhaps we've only seen such people while they proselytized their newfound faiths in the airports or streets. Regardless, the fact that this psychological phenomenon occurs to a certain percentage of every population and in every religion implies that it represents yet another integral aspect of our species' inherent neural processing.

Just as we are the musical, emotional, and linguistic animal, we are also the "converting" animal, the animal whose sense of personal identity can be suddenly and drastically transformed in such a way that religious concerns come to predominate conscious experience. In his book *Varieties of Religious Experience*, William James was one of the first to document this uniquely human behavior. As James expressed it, "to say that a man is 'converted' means that religious ideas, peripheral in his consciousness, now take a central place, and that religious aims form the habitual center of his energy."[89]

For people who undergo religious conversions, individuality is replaced by ideology, and very little room is left for personal growth or expression. Because the converted believes that his newfound faith determines all things, all sense of personal responsibility is relegated to some religious credo or God. To the converted, all that occurs does so because God willed it such. No matter how ordinary or mundane an occurrence might seem, God is suddenly viewed as responsible for everything. Should something unfavorable occur, it is because "God works in mysterious ways." Tragic events become "blessings in disguise" as all events, good or bad, are smoothed over by a feeling of being unconditionally loved by a higher power, what John Wesley, the founder of Methodism, referred to during his own conversion as a feeling of "the heart strangely warmed." Is it possible that, as has been demonstrated with meditative, mystical, or near-death experiences (chapter fifteen), such sensations have a direct correlation to changes in one's neurophysiology?

Concerning the conversion process itself, though some take place in a slow and gradual manner, the majority of cases occur very abruptly. Many psychologists, such as E. S. Ames, favored "restricting the term 'conversion' to sudden instances of religious change."[90] G. A. Coe also thought that the use of the term conversion should be limited to those cases in which the individual undergoes an intense and sudden religious change. The only other time a person's core personality undergoes such an abrupt and drastic change is when stricken by an organic psycho-syndrome or psychosis. Being that both of these are listed as disorders in the DSM-IV, one has to wonder why we don't view religious conversion in the same negative light (i.e., as a pathological condition as opposed to a "blessing").

In studying the etiology of religious conversion, we need to look at those triggers that seem to precipitate the experience. According to the psychologist Paul Johnson, "A genuine religious conversion usually occurs as the outcome of a crisis of ultimate concern and a sense of desperate conflict."[91] In his book *Religious Conversion: A Bio-psychological Study*, the psychologist S. De Sanctis asserts that "all the converted speak of their crisis, of their efforts, and of their conflicts which they have endured."[92] In his work *The Cognitive and Emotional Antecedents of Religious Conversion*, C. Ullman discusses studies he conducted in which he compared psychological traits of those who had undergone a religious conversion and those who had not. Ullman found that:

> Converts recalled childhoods that were less happy and filled with more anguish than those of non-converts. The emotions recalled for adolescence followed similar childhood patterns, with the addition of significant anger and fear in adolescence for the converts and not the nonconverts. Converts also differed from the unconverted in having less love and admiration for their fathers and more indifference and anger towards them.[93]

After studying 2,174 cases of religious conversions, the psychologist E. T. Clark noted, "Sudden conversions were associated with fear and anxiety."[94]

If we were to look for a pattern, it would seem that those most susceptible to this type of sudden cognitive transformation are those with fragile senses of identity and unhealthily developed egos, those who were abused or neglected by their parents, without whose love, they were never able to feel secure in the world. When such a child grows, he may not possess the inner strength and personal stability required to endure life's ordinary trials and tribulations, thus catapulting that person into a state of emotional crisis. When their crisis reaches a threshold, a breakdown occurs in which the suffering individual latches on to some religion to which he will soon convert.

Follow-up studies show that after such deeply troubled individuals undergo their religious conversion, their emotional states generally tend to improve. According to a study done by J. B. Pratt, "Prior to conversions, individuals had a tendency to wallow in feelings of unworthiness, self-doubt, and depreciation that are released or overcome via conversion."[95] Yet another study showed that "it is typical of conversion to be preceded by morbid feelings in which doubt, anxiety, internal strife, and despair are replaced by serenity, peace, and optimism."[96] Apparently, for those who suffer severe emotional turmoil, there are obvious benefits in undergoing a religious conversion.

Consequently, many religious groups intentionally seek out the lonely and afflicted because they know they are most likely to succumb to a conversion. The theologian Lewis Rambo points out that certain religious groups such as the Evangelical Christians make it part of their practice to target vulnerable individuals. For example, in large urban areas some churches focus on ministries to those recently divorced as they know that within the first six months after a divorce, a person is more likely to be converted. This practice of seeking out those in crisis is most evident among prison populations, where stress levels are critical and conversions are practically endemic. Another example in which the vulnerable are targeted for conversion is practiced by recovery groups such as Alcoholics, Eaters, Gamblers, and Debtors Anonymous, all of which emphasize–through the use of the renowned "12 step" program–faith in religion and God as primary tools in their effort to combat these addictive behaviors. When one is trying to overcome an addiction, they experience a sharp rise in stress levels, rendering them prime candidates for a religious conversion.

Because a percentage of every population undergoes this type of sudden behavioral change, it suggests that religious conversions most likely constitute yet another inherent characteristic of our species, a genetically inherited reflex response to overwhelming crisis or anxiety. If this does, in fact, constitute a physiological reflex, it would suggest that there must exist some specific neurophysiological mechanism responsible for generating this type of behavioral response.

It appears that the human capacity to endure existence is so tenuous that nature had to install our species with an emergency back-up identity–a religious one–to replace our secular ones when they can no longer provide adequate relief from our excess anxieties. Depending on to what degree one is genetically predisposed to undergo a conversion, we each have our own personal threshold for pain and duress that we can withstand before we, too, become vulnerable to undergoing this type of cognitive transformation. It appears that when a person reaches his own personal anxiety threshold, rather than to engage in some self-destructive behavior, such as drug abuse or some other addictive behavior through which to mask our pain, one's "normal," secular self shuts down and is immediately replaced with an alternate hyperreligious one. Once the convert's cognitive transformation is complete, he is relieved of all those anxieties that were attached to his previous identity. All past fears and anxieties are washed away and replaced with rapturous contentment and a feeling of being safe and secure. No wonder many converts refer to themselves as being "saved."

The human ego is a very delicate organ. If it is not properly nurtured, a person may grow to develop any number of insecurities, neuroses, or even psychoses. When a person with a weak sense of self reaches the preliminary stages of adulthood, he or she may not feel ready or able to take on life's responsibilities. Perhaps this is why religious conversions "typically occur during adolescence."[97] This is further supported by the research of psychologist Paul Johnson who concluded that "after surveying five studies conducted on over 15,000 people, the average age of conversion was 15.2 years."[98]

This is not to suggest that religious conversion only occurs during adolescence, for it can strike at any age that a person feels particularly vulnerable and/or threatened. Nevertheless, it is during adolescence that humans are generally afflicted with increased anxiety levels as it is during this age that we are first told by our parents as well as society that we're soon going to have to fend for and support ourselves. Moreover, it is during adolescence that we must first come to terms with the concept of our own mortality.

With all of these concerns, questions, pressures, and responsibilities suddenly thrust upon us, it is no surprise that it is during this same stage in our developments, usually between the ages of fifteen to twenty, that humans undergo the most cases not only of religious conversion, but of suicide, drug abuse, eating disorders, depression, and schizophrenia. It is therefore also no wonder that the majority of conversions take place at this same age, as research suggests that increased religiosity can lead to a reduction in a number of self-destructive behaviors. Regarding that most self-destructive act of all–suicide–W. T. Martin, in his 1984 article titled "Religiosity and United States Suicide Rates," reported that "church attendance remains negatively correlated with suicide rates."

This was further supported by research done by H. G. Koenig, who concluded in his work *Aging and God* that among the elderly "faith suppresses suicidal thinking." After interviewing a number of individuals, Koenig found that many expressed that "the promise of a happy afterlife" helped to thwart any suicidal inclinations. In another study, the team of S. Stack and I. Wasserman found that a belief in an afterlife helped to counter self-destructive impulses. Apparently, those who do not believe in a spiritual reality are more prone to engaging in self-destructive behaviors than those who have faith. Perhaps it's for reasons such as this that, although we are well aware of the radical personality changes caused by religious conversion, we are reluctant to classify it as a psychological disorder. At the same time, however, it's worth noting that although we are fairly accepting of individuals who convert to mainstream religion, when a person converts to an unsanctioned one–what is otherwise referred to as a cult–it's much more discouraged, often leading families or societies to intervene by trying to seize the convert from the clutches of what is viewed as an insidious group or influence. Regardless of how we choose to perceive this strictly human phenomenon, we must accept that it represents another cross-cultural characteristic of our species and therefore, more than likely, another genetically inherited predisposition, reconfirming the notion that human spirituality and religiosity are products of our biologies and not of some mystical influence or God.

Chapter 14

Why Are There Atheists?

"What is going to happen to those of us who want to believe but aren't able to? And what is to become of those who neither want to nor are capable of believing?"

—Ingmar Bergman, *The Seventh Seal*

In discussing the essential precepts of biotheology with others, one of the questions most frequently asked has been, "If human spirituality represents an inherent characteristic of our species, if we truly are 'wired' to believe in a spiritual realm, in a God, then why are there atheists?" In essence, if we, as a species, are "wired" to believe in such things, how do we explain those who don't?

Though we may all exist as part of the same species, no two human beings are exactly alike. As similar as we might be, each of us is a unique composite of physical and cognitive traits. Whereas some of us are taller than average, others are shorter. While some have exceptional vision, others are born blind. Whereas some are more musically or mathematically gifted, others are born deficient in these areas. As a matter of fact, the distribution of every genetically inherited trait can be charted by a bell curve.

To demonstrate, let's apply this notion to a basic physical characteristic such as vision. Though the majority of our species is born with average eyesight and will therefore fall somewhere within the mean of this curve–within its bulge–there exists a much smaller percentage of individuals within every population that represent the tapering ends. Whereas one end of this curve is represented by those born with superior vision, on the opposite side there will most probably exist an equally small number of individuals who are born with inferior vision, with some on the extreme edge who are totally blind.

Just as this precept can be applied to any inherent physical trait, it applies to cognitive traits as well. Take musical ability, for example. Though most of us are born with an "average" capacity to develop certain musical skills from composing to playing an instrument, each population possesses a smaller percentage of individuals who fall into one of this curve's two tapering ends. On one side, every culture possesses a minority of those born musically gifted. At this end's extreme, there exist an even smaller number of exceptionally gifted (i.e., savants, such as Mozart). Meanwhile, on the opposite extreme of this same curve, each population will most likely possess an equally small percentage of those born musically deficient–or in some cases even tone deaf–who, though they can hear, lack any inherent musical intelligence and who don't even have the capacity to learn musical skills.

For every capacity we possess, cognitive or otherwise, there must exist a physiological site from which that capacity is generated. Our capacity for vision, for instance, is directly related to the caliber of our eyes and visual cortex. Similarly, our capacity for music is directly related to the caliber of those parts of the brain responsible for generating musical ability. We could therefore say that whereas someone like Mozart must have been born with an unusually overdeveloped musical part of his brain, someone less gifted is most probably born with a less developed part of theirs.

This, of course, is not to exclude the environmental factor. Though each of us is born with a certain degree of inherent potential in any number of abilities, the degree to which we actualize those latent capacities depends on to what degree we nurture and cultivate them. Had I, for instance, been provided with a great deal of musical instruction from early childhood on, I'm sure I would possess a greater degree of musical ability than I do today. Nevertheless, even with the most intensive musical training conceivable, there's no way I could have ever matched Mozart's skills simply because I was not born with the same genetic potential to achieve his level of skill.

The same holds true for the opposite scenario. Mozart, for instance, were he born to peasants, indentured to till the soil, without the same opportunity to study music as he had, would never have reached the level of genius he achieved in his lifetime. In such a case, he may have instead merely grown to become "the guy who whistles really well while toiling the field." Unfortunately, in the same vein, latent Mozarts, Einsteins, and Michelangelos probably die every day without the slightest recognition simply because they were never afforded the opportunity to actualize their inherent genetic potentials. I'm therefore suggesting that while life experience (nurture) plays a significant role in our cognitive developments, we can only reach as high as our inherent genetic potentials (nature) permit.

So what does any of this have to do with the question of atheism? Since it appears that both spirituality and religiosity are generated from specific regions within the brain, mustn't the aforementioned "bell curve" principle apply to these inherent proclivities as well? If we, in fact, do possess neurophysiologically based spiritual and religious mechanisms, then wouldn't it make sense that the average person from any given population would probably possess an average potential for either of these intelligences?* Regarding these same bell curves' tapering extremes, every population should therefore also possess a smaller percentage of those born with either an enhanced or diminished capacity for either of these two distinct cognitive traits.

Regarding those who fall into the mean of the spiritual/religious curve, such individuals are likely to possess enough spiritual/religious intelligence that they will be predisposed to believing in some form of a transcendental reality. These are our masses, the bulge of the spiritual bell curve, those who have kept spiritual ideals along with religious institutions thriving for all these years as an integral part of every human society.

Regarding this trait's tapering extremes, on one end of this curve are those born with an overdeveloped spiritual/religious function, those for whom spirituality/religiosity will play a predominant role in their conscious experience. On the farthest extreme are those who, even as early adolescents, will be delivering heartfelt sermons from the pulpit, those of whom we might say were "born with the spirit in them." These often turn out to be our prophets, zealots, mystics, fundamentalists, martyrs, and spiritual leaders, those born with a greater predisposition toward hyperreligiosity, or what we might call an overdeveloped spiritio-religious function.

On the opposite end of this same curve there are those we might call spiritually/religiously deficient, those born with an unusually underdeveloped spiritual/religious function. Just as a person born blind is light-insensitive, those born with an underdeveloped

*To reassert the distinction between spirituality and religiosity, we must realize that though they usually operate in tandem with one another, one can still be born with any combination of these two unique impulses. For instance, though one might be born with an underdeveloped religious impulse, he might have an overdeveloped spiritual one. Though such an individual might not be inclined to attend church or engage in religious rituals, he might be very spiritual, highly prone to undergoing "transcendental" experiences. On the other hand, there are those who are hyperreligious, though aspiritual. Such individuals, though they may never have a spiritual/mystical experience or feel compelled to contemplate any "higher" truth or reality, might be obsessed with the rigid adherence of church doctrine, custom, and code. It is these individuals who are most prone to the dangerous excesses of religious fanaticism.

spiritual function are spiritually insensitive, incapable of fully grasping, appreciating, or experiencing the concept or implications of any spiritual reality. Such people rarely, if ever, feel compelled to worship or pray, to consider or contemplate the concepts of a spiritual reality, a god, a soul, or an afterlife. Such people are unlikely to ever have a spiritual experience. These are society's spiritually retarded, if you will, or, in keeping with the musical metaphor, those we might call spiritually tone deaf. Just as a person can be born mathematically or musically deficient, it is just as likely that a person can be born spiritually or religiously deficient. It is here that we will find the neurophysiological origins of those with a greater predisposition toward agnosticism and atheism, our rationalists and secularists.

To again account for the environmental factor, we must realize that atheism is not exclusively dependent upon one's genes. In many cases atheists are those who, though they might be inherently spiritual, were raised in a nonreligious or aspiritual environment, in which case their innate proclivities may have atrophied and consequently been substituted by a secular world view. At the same time, there are also those who, though inherently spiritual, have become so disenchanted with organized religion that they've chosen to suppress their inherent proclivities and consequently deny a belief in any religion or God.*

---

*As most atheistic ideologies are based in the mere denial of God's existence, I would like to stress that no philosophy can be justifiably upheld without possessing some underlying logic through which to substantiate its basic principles. Without such a logic, what is referred to as a philosophy is really nothing more than just another groundless belief system, founded in emotion rather than reason. As I see it, this is the essential problem faced by today's atheist movement. Rather than possessing an inherent wisdom of its own, the atheist movement relies on the logical shortcomings of those faiths it seeks to contest. And though it's true that no religion has ever been able to defend its precepts with reason, no legitimate philosophy can stand on gainsay alone. The contradicting of one belief system does not validate the tenets of another. Establishing that something is not white, for instance, does not necessitate its being black. Analogously, finding fault in the convictions of every world religion does not constitute proof that there is no God. Consequently, if we are ever to advance a viable atheism, it must possess its own rationale, its own logical foundation, something I believe this new science of "biotheology" finally provides.

CHAPTER 15

NEAR-DEATH EXPERIENCES

"Mysteries are not necessarily miracles."

—GOETHE

We are all, to some extent, familiar with the phenomenon known as a near-death experience. Whether we have had such an experience ourselves or have merely heard of one as recounted by a friend or a guest on some TV talk show, the near-death experience (NDE) has been reported by a cross-section of every world population and must therefore constitute another inherent part of the human cognitive experience. As with all other cross-cultural behaviors, this would suggest that the NDE most likely represents the consequence of a genetically inherited trait, a biologically based response, a reflex to a specific stimuli. Though NDEs are generally interpreted as "spiritual" in nature, the result of an encounter with the afterworld, as with all other types of spiritual experiences, I believe that these, too, are no more than the effects of strictly neurophysiological processes.

References to NDEs date back to Plato's *Republic* as well as *The Tibetan Book of the Dead* and have been reported by nearly every world culture since. In his *Republic*, Plato recounts the tale of Er, the son of Armenius, who allegedly dies and then comes back to tell the story of his temporary ascension to heaven and consequent return to the living. During Er's alleged experience with death, he describes a vision he had of a "bright and pure column of light, extending right through the whole of heaven." It is through Er's tale that Plato goes on to advance his notion of an immortal soul as well as an afterlife. In this way, NDEs tend to play a significant role in reinforcing our species' beliefs in a spiritual reality and an afterlife.

In order to explore the frequency of NDEs in more contemporary societies, the Gallup organization published a national survey in 1982 called "Adventures in Immortality" that set out to examine what adult Americans believe about life after death. One of the questions asked was, "Have you yourself ever been at the verge of death which involved any unusual experience?" In response, 15 percent said they had. Moreover, in this same poll, it was surmised that as many as eight million North Americans have had an NDE. In a similar survey conducted in China (Feng and Lin, 1976), 42 percent of those questioned claimed to have undergone an NDE, lending support to the cross-cultural nature of this phenomenon.

And what precipitates an NDE? NDEs almost always occur as a result of decreased blood flow to the brain and/or lack of oxygen, usually from shock induced either from severe infection (septic shock), from myocardial ischemia (cardiogenic shock), cardiac arrest, or the effects of anesthesia. Apparently, NDEs are integrally linked to physical–not spiritual–realities.

One of the most common misperceptions regarding NDEs is that when we have one we literally die and are then restored to life, something which is simply not possible. For example, some people mistakenly believe that when our heart has stopped, we are dead. Contrarily, the heart is merely a pump that sends oxygenated blood to the rest of the body. It is not until approximately six minutes after a cell has been deprived of its normal oxygen supply that it truly dies. Not until the cells in a person's brain have died are we truly deceased, a death from which no living organism has ever returned.

Though there is no international standard through which to formally define an NDE, studies show vast similarities in descriptions of this phenomenon, ones that cross all cultural boundaries (Fenwick, 1997; Feng and Lin, 1976; Parischa and Stevenson, 1986). For example, in the majority of recorded accounts, the first thing most people recall of their experience is a feeling of intense fear and pain that is abruptly replaced by a sense of calm, serenity, and bliss. To offer support of a neurophysical model of this phenomenon, D. B. Carr suggested that the aforementioned sensations, in so far as they are experienced during an NDE, might come as the result of a flood release of endogenous opiods (endorphins).

After experiencing this sense of calm or euphoria, the next most often related symptom to occur during an NDE is that of an "out-of-body" experience (OBE). Here, the person describes a sensation of having risen outside of their physical body and, in many cases, even being able to look down at one's own self from above.* During this part of the experience, those undergoing an OBE have expressed a sense that their limbs were "moving within their mind," though, as the doctors within the room can confirm, they were completely immobile. This is similar to the type of hallucinations, or confabulations, suffered by those who sustain right parietal lesions, yet another indication that such experiences can be traced to neurophysical activity as opposed to originating from one's alleged spiritual self or soul.

Another common symptom of the NDE, similar to the one narrated by Plato, is described as a sensation of being led down a dark tunnel and then drawn toward a blinding white light, one that is often interpreted as holding religious significance, such as being representative of heaven's gates. Such descriptions as these, of experiencing a piercing or dazzling white light, have been attributed to activity within the brain's optic nerve which has a tendency

*One hospital, in order to validate claims of "out-of-body" experiences, placed an LED marquee above its patients' beds which displayed a hidden message that could only be read if one were looking down from above. To date, not one person who has claimed to have had an NDE or OBE within that hospital has expressed having seen the hidden message.

to react when deprived of normal oxygen supplies. It is during this same part of the experience that a person will often express a feeling of being engulfed, not just by "the light," but by the presence of God.

Similar to accounts of those who have undergone temporal lobe epilepsy or experimented with entheogenic drugs, those who have had an NDE almost invariably interpret their experiences as being spiritual in nature. As reported in the *Journal of Neuropsychiatry*:

> Hallucinogen ingestion and temporolimbic epilepsy produce a near identical experience as is described by persons having a near-death experience. These brain disturbances produce depersonalization, derealization, ecstasy, a sense of timelessness and spacelessness, and other experiences that foster religious-numinous interpretation.[99]

Consequently, it is no surprise that a significant number of those who undergo an NDE claim that it strengthens their belief in a spiritual reality, a god, a soul, and an afterlife. Nevertheless, regardless of how we choose to interpret these experiences, we must ask ourselves: Is this type of experience transcendental in nature or, like all other types of spiritual experiences, are we dealing with a series of strictly neurophysical events?

One key to answering this question comes through the research of a Dr. Karl Jansen, who has found that "near-death experiences can be induced by using the dissociative drug ketamine."[100] Dr. Jansen's report goes on to state that, "It is now clear that NDEs are due to the blockade of brain receptors (drug binding sites) for the neurotransmitter glutamate. These binding sites are called the N-methyl-D-aspartate (NMDA) receptors. Conditions which precipitate NDEs, (i.e., low oxygen, low blood flow, low blood sugar) have been shown to release a flood of glutamate, over-activating NMDA receptors. Conditions which trigger a glutamate flood may also trigger a flood of ketamine-like brain chemicals, leading to an altered state of consciousness."[101]

It was also found that an intravenous injection of 50–100 mg of ketamine reproduces all of the symptoms commonly associated with a near-death experience (Sputz, 1989; Jansen, 1995, 1996). Even Timothy Leary, the notorious psychedelic drug advocate of the 1960s, described his experiences with ketamine as an "experiment in voluntary death" (Leary, 1983).

Given that NDEs occur, as the name suggests, when our lives are at stake, it would make sense that the body would release chemicals that induce a state of calm and serenity. For instance, if we are in the process of bleeding to death, the worst thing we can do is to panic, which will only increase our heart rates, which would only expedite the rate of blood loss. Rather, it's to our advantage that the body should induce a state of calm and euphoria that will slow our heart rates, thus decreasing the rate of blood loss. This is most likely the adaptive function of an NDE–to calm us in the midst of life-threatening events so as to bolster our chances of survival.

Similar to the manner in which entheogenic drugs can trigger a spiritual/mystical experience in us, it appears that the neurotransmitter glutamate as well as its synthetic substitute, ketamine, can induce all of the symptoms of a near-death experience. What this suggests is that, similar to all other types of spiritual experiences, NDEs are rooted–not in any ethereal soul, but in our brain's neurochemistry. Apparently, the NDE represents the consequence of a neurophysiological mechanism that enables our species to cope with the overwhelming pain and anxiety associated with a life-threatening circumstance. Once again, although such physical evidence can never prove that no spiritual reality exists, it certainly bolsters the possibility that this may very well be the case.

Chapter 16

Speaking in Tongues

"Glossolalia is a universal religious phenomenon."
–C. L. May

Another religious behavior that warrants addressing is that of glossolalia, also referred to as ecstatic experiences, or more commonly known as "speaking in tongues." Glossolalia constitutes yet another universally enacted behavior through which the human animal can induce a trancelike state very similar in its description to the types of spiritual/mystical experiences discussed in chapter three.

To confirm the cross-cultural nature of glossolalia, the ethnologist George Jennings studied this strictly human phenomenon as experienced by a variety of world cultures which include the peyote cult of the North American Indians, the Haida Indians of the Pacific Northwest, shamans in the Sudan, the Shango cults of the west coast of Africa and Trinidad, the Voodoo cult in Haiti, the Australian Aborigines, the aboriginal peoples of the subarctic regions of North America, the shamans in Greenland, the Dyaks of Borneo, the Zor cult of Ethiopia, the Siberian shamans, the Chaco Indians of South

America, the Curanderos of the Andes, the Kinka in the African Sudan, and the Thonga shamans of Africa.

Among Christian societies, glossolalia can be traced back to the writings of the New Testament (Acts 2:1–42) in which Paul and Luke indicate that speaking in tongues was a notable part of the early Christian church. According to these writings, speaking in tongues was regarded as the effect of the Holy Spirit taking possession of a Christian's body.

As is true of many spiritual/mystical experiences, glossolalia is generally evoked in a formal religious setting. Among the Pentecostal Christians, for instance, special "revival" meetings are held meant to produce an atmosphere that will encourage the participants to engage in this type of ecstatic experience. Like the Islamic whirling dervishes who spin themselves into an ecstatic frenzy, an individual seeking to evoke glossolalia must work himself into a similar religious fervor. Once this "heightened" state is reached, the initiate will involuntarily engage in a series of unintelligible utterances, word fragments, and vocalizations from which the phenomenon derives its name. Like other types of spiritual/mystical experiences, those who practice glossolalia often describe their experiences as producing feelings of religious rapture and ecstasy.

So, are we to believe that such experiences represent genuine instances of humans being possessed by a spirit? Are the unintelligible utterances that come as a result of speaking in tongues really the vocalizations of our gods who are merely using us as their mouthpieces? Or is it possible that here lies yet another neurophysiologically based human reflex?

Though little is yet known of the underlying biology of the experience of glossolalia, through the use of electroencephalographic (EEG) recordings, it was revealed that a distinct change occurs in the brain wave patterns of those entering what the participants referred to as an "anointed" state of consciousness (Woodruff, 1993). More specifically, it was found that as the participants entered this heightened state, their brain wave patterns suddenly shifted from alpha to beta, thus confirming that such experiences have direct correlation to neurological activity.

The physical connection between this type of religious experience and our neurophysiology was further validated in experiments conducted by V. S. Ramachandran and S. Blakeslee in 1998 that showed that the right cerebral hemisphere plays a major role in glossolalia. In addition, experiments conducted on glossolalia subjects which revealed a temperature change in the right and left hemispheres also suggest that "the experience of speaking in tongues may be associated with increased activation of the right hemisphere."[102]

Here lies another example of a human experience that, though it is often conceived as being "spiritual" in nature, we are finding can be traced with the aid of science not to some divine source, but rather to activity being generated from within the human brain.

Chapter 17

Why Is America So Religious? A Bio-Historical Hypothesis

"We the people of the United States now form the most profusely religious nation on earth."[103]

–Diana Eck

"We are a religious people whose institutions presuppose a Supreme Being."

–Supreme Court Justice William O. Douglas[104]

"In God We Trust"

–U.S. Currency

According to recent demographic studies and social statistics, there appears to exist an inverse relationship between a nation's prosperity and the extent of its religiosity. In other words, whereas the more prosperous nations of earth possess a statistically higher percentage of those who define themselves as being non-religious, atheistic, agnostic, secular, or unaffiliated with any faith, the least prosperous nations possess significantly higher numbers of those who define themselves as being religious.

To confirm this correlation, the Human Development Report of 2004, which was commissioned by the United Nations Development Program, ranks 177 nations on what they refer to as the "Human Development Index." The point of this index is to measure a nation's societal health by utilizing such indicators as infant mortality rate, adult literacy rate, per capita income, and educational attainment. According to the 2004 Report, the five nations that ranked highest on this scale were Sweden, Norway, Australia, Canada, and the Netherlands. Not only are all five of these nations characterized by notably high degrees of secular atheism but "of the top 25 nations ranked on the Human Development Index, all but one country (Ireland) are top-ranking nonreligious nations, containing some of the highest percentages of atheism on earth. Conversely, of those countries ranked at the bottom of the Human Development Index–the bottom 50–all are countries lacking any statistically significant percentages of atheism."[105] Moreover, "Concerning literacy rates, according to the United Nations' Report on the World Social Situation (2003), of the 35 nations with the highest levels of youth illiteracy rates (percentage of population ages 15–24 who cannot read or write), all are highly religious nations with statistically insignificant levels of atheism."[106]

In addition, whereas nations that are statistically less religious possess greater levels of gender equality and are among the most egalitarian in the world, nations that are more religious are considered among the most oppressive with high degrees of gender inequality.

To offer a few statistics which support these claims: 42 percent of West Germans and 72 percent of East Germans are either atheist or agnostic (Shand, 1998), 85 percent of Swedes do not believe in God (Davie, 1999), 44 percent of the British do not believe in God (BBC, 2004), 65 percent of those in Japan do not believe in God (Norris and Inglehart, 2004), 54 percent of the French are atheist or agnostic, 44 percent of the Dutch are either atheist or agnostic (Froese, 2001), while 54 percent of Israelis identify themselves as secular (Yuchtman-Ya'ar,

2003), 31 percent of Israelis do not believe in God, with an additional 6 percent choosing "don't know," for a total of 37 percent being atheist or agnostic (Kedem, 1995). Of these countries, all but Israel are listed in the UN's 2005 Human Development Index among the top twenty "most livable countries" in the world.

Conversely, in the majority of countries in the Middle East, Asia, South America, and Africa, less than 1 to 2 percent of people are either nonreligious or do not believe in God.*

And now for the question I've so conspicuously saved for last: Where does America fit in all this? What are our numbers? Moreover, are they consistent with the statistical correlations obtained from the surveys of all these other nations? The short answer: not even close. We are a glaring anomaly.

According to a Gallup Poll released in November 2003, 60 percent of Americans said that religion was "very important" in their lives. Meanwhile, in Canada and the United Kingdom, two nations with whom we consider ourselves sharing the greatest cultural affinity, only 28 percent and 17 percent similarly defined religion as being important in their lives. A survey conducted by the City University of New York Graduate Center in 2001 found that 85 percent of Americans identify with some religious faith.

According to the British-based polling agency ICM, a survey taken in January 2004 found that 91 percent of Americans believe

---

*Among those nations polled in which only 1–2 percent of their populations consider themselves non-religious or atheistic are Jordan, Egypt, Syria, Oman, Kuwait, Saudi Arabia, United Arab Emirates, Iraq, and Yemen (Inglehart et al, 2004; Barret et al, 2001), Indonesia, Bangladesh, Brunei, Thailand, Sri Lanka, Iran, Malaysia, Nepal, Laos, Afghanistan, Pakistan, and the Philippines (Gallup, 1999; Johnstone, 2003), El Salvador, Guatemala, Bolivia, Brazil, Costa Rica, Colombia, Ecuador, Honduras, Nicaragua, Panama, Peru, Paraguay, and Venezuela (Hiorth, 2003; Barret et al, 2001; Inglehart et al, 2004), Algeria, Benin, Botswana, Burkina Faso, Burundi, Cameroon, Chad, Cote D'Ivoire, Ethiopia, Gambia, Ghana, Guinea, Kenya, Liberia, Libya, Madagascar, Malawi, Mali, Mauritania, Morocco, Niger, Nigeria, Rwanda, Senegal, Sierra Leone, Somalia, Sudan, Tanzania, Togo, Tunisia, Uganda, Zimbabwe, and Zambia (Hiorth, 2001; Inglehart et al, 2004, 1998; Barrett et al, 2001; and Johnstone, 1993).

in the supernatural, 74 percent in an afterlife, 82 percent think that believing in a God/higher power makes you "a better human being," and 76 percent believe a God or a higher power judges their actions, while 71 percent claimed that they "would die for their God/beliefs." In contrast, only 5 percent of Americans do not believe in God or a higher power (Gallup, 1999). Moreover, based on an ABC news poll done in February 2004, 60 percent of Americans believe in a literal interpretation of such biblical accounts as the Genesis creation, the parting of the Red Sea, and the story of Noah's ark.

Clearly, America is a significantly religious nation. As a matter of fact, of the top fifty countries containing the largest percentage of people who identify themselves as non-religious, America ranked forty-fourth, following such underdeveloped nations as Uruguay, Kazakhstan, Estonia, and Mongolia. Moreover, whereas none of these above mentioned nations placed anywhere near the Human Development Index's top ten most livable nations, America did.

So why this striking disparity? How is it possible that in a nation as prosperous as the United States–one with such a low infant mortality rate, high adult literacy rate*, high degree of gender equality, high per capita incomes and standards of living, the inventor of so many successful technologies, a society so immersed in scientific culture, not to mention winner of the most Nobel prizes in science (possessing more than all of the next five leading recipient nations combined between 1901 and 2003: America, 137; Germany, 49; United Kingdom, 47; France, 18; Netherlands, 11; Russia, 11)–is so characteristically religious? Such findings beg the question: What is it about America that it so defies those same social statistics that consistently resonate with nearly every other nation in the world?

In keeping with this book's biotheological premise, I would like to offer a genetic hypothesis of this apparent phenomenon: To speak of a people's national character, one need first look at the origins of

*The adult literacy rate of the U.S.A. is 97 percent, which, though it's still a relatively high percentage, is still lower than almost all of the other developed nations.

that people they are seeking to characterize. In the case of America, its origins lie in the tale of seventeenth century European immigrants. And why did these people leave Europe to settle here? Though they came for an assortment of reasons, economic prosperity among them, if there's any one common denominator by which we could distinguish nearly all of the first European communities to settle on North America, it could be summarized in that they came seeking religious freedom–men and women whose religious convictions were so strong that they were willing to risk life and limb in order to practice their faiths as they saw fit.

The first of these religious groups to settle on North America were the Pilgrims. During the sixteenth and seventeenth centuries, a religious group known as the Puritans were seeking to purify (from which they derived their name) the Anglican church of England by reforming its policies and divesting it of all vestiges of Rome and its popery. From an offshoot of this group, an even more radically separatist sect emerged, many of whom left England to live in Leyden, Holland to practice their brand of Christianity. Of this Leyden congregation, a group known as the Pilgrims decided to establish their own community in the New World, where, in 1620, they founded the colony of Plymouth.

In the face of growing persecution within Anglican England, a second wave of Puritans fled to North America and founded the Massachusetts Bay Colony in 1630. This time as many as twenty thousand Puritans emigrated to America from England with the sole purpose of freely practicing their religion.

In light of the success of these first Puritan communities, other religious dissidents soon followed their lead. One of the first of these other groups were the Huguenots, a society of French Protestants. At the time atrocities were being committed by both sides as Catholics and Protestants waged a war that engulfed much of Europe. One of the worst of these incidents took place during the Bartholomew's Day Massacre in Paris (1572) in which thousands of Huguenots were slain by a Roman Catholic mob. Although a tentative peace was declared in the Edict of Nantes in 1598, tension between these two groups eventually compelled the Huguenots to seek greener pastures, inspiring as many as four hundred thousand of them to emigrate to various parts of the world, including the British North American colonies.

During the reign of the Stuart kings of England in the seventeenth century, Catholics were being consistently harassed and persecuted. Driven by a feeling of sacred duty, George Calvert, a member of the British Parliament who had converted from Anglican Protestantism to Catholicism, obtained a charter from Charles I in 1632 for the territory between Pennsylvania and Virginia or what became known as the Maryland Charter. Calvert offered those willing to join him the freedom to practice their faith with impunity, and, in 1634, two ships, the Ark and the Dove, brought their first settlers to Maryland where they set up the first Roman Catholic communities in the New World.

For years, Dutch Jews who had flourished in Dutch-held areas of Brazil were suddenly confronted with the threat of inquisition after a Portuguese conquest of the area in 1654. After one Brazilian Jew had already been burnt at the stake, a shipload of twenty-three Jewish refugees from Dutch Brazil took flight to New Amsterdam (what would soon become New York) to find religious asylum. It was here in New Amsterdam that these men and women established the first Jewish congregation on North America as well as erected the first synagogue. This represented the first of three separate waves of Jewish emigration to America. In the second wave, which took place in the nineteenth century, a large community of German Jews left that country for a better life in America. During the third wave, which was also the largest, Jews fled from Poland and Russia between 1881 and 1906 to escape extreme persecutions known as pogroms that were authorized by the czar.

In 1652 a religious community had arisen in England around a charismatic leader named George Fox, who founded the Quaker movement. The Quakers, who at the time were viewed as radical Puritans, were severely persecuted in England for daring to deviate from orthodox Christianity. By 1680, the nation of England had imprisoned over ten thousand Quakers, some of whom were tortured to death in the king's jails. As a result, the Quakers sought refuge in the New World where Quaker leader William Penn had secured a charter from Charles II for the province of Pennsylvania. By 1685, as many as eight thousand Quakers had established communities in Pennsylvania.

As a result of the Thirty Years' War (1618–1648), many of Germany's Protestant communities–namely Mennonites, Dunkers, Schwenkfelders, and Moravians–found themselves victims of religious persecution. On hearing this, William Penn, who had by now established his Quaker community in Pennsylvania, began circulating literature to these oppressed German religious groups touting the advantages of living in Pennsylvania and encouraging them to join him there. In response, thousands of these Germans sailed to the New World, where they found religious freedom in Pennsylvania. As a result of this mass immigration of so many different religious groups, the province became what one author described as "an asylum for banished sects."

The New England communities were referred to as "Bible Commonwealths" in that they were virtual theocracies through which biblical scripture was to be interpreted as societal law. By 1609, the Church of England had been established as law in Virginia, and in 1610 a statute was added that made church attendance compulsory. This trend continued as Anglican law was soon extended to New York in 1693, Maryland in 1702, South Carolina in 1706, North Carolina in 1711, and Georgia in 1758, with the rest of the colonies following thereafter. With such an influx of predominantly religiously oriented populations, by 1700 it was estimated that between 75 to 80 percent of the colonies' populations regularly attended church, of which new ones were being built at an expeditious pace.

Before I posit any conclusion regarding the aforementioned data, I'd like to offer an analogy: imagine we were to take the entire New York Philharmonic–let's say a couple hundred people altogether, people not only possessing a distinct passion for music but also a heightened degree of inherent talent–and we were to banish them to an isolated island. Now imagine two hundred years later we were to pay a visit to their surviving progeny: Would it be unreasonable to presume that we would most likely find a society steeped in musical culture? Granted, as the island's founders would most likely have stressed musical appreciation and education to their offspring, a great deal of this could be attributed to environmental factors. Nevertheless, isn't it also reasonable to presume that some portion of this societies' musical nature might result from inherent aptitudes and proclivities passed on by their forefathers' enhanced musical genes? Even if several generations into this island societies' genesis new immigrants were to arrive–many with little or no inherent musical talent or inclination–isn't it highly probable that the island's strong musical heritage would still persist to some extent?

Such a case would represent a hypothetical example of the "founder" or "pioneer effect," that facet of the evolutionary process known as genetic drift in which a small group from a much larger population migrates to an isolated area, bringing with it a unique genetic admixture from which, generations later, entirely new species can emerge, or, as in the case with humans, new races or cultures possessing unique physical features and possibly even genetically inherited behavioral predispositions.

To provide an actual example of this force at work, "The founder effect is probably responsible for the virtually complete lack of blood group B in American Indians, whose ancestors arrived in very small numbers across the Bering Strait during the end of the last Ice Age, about ten thousand years ago. More recent examples are seen in religious isolates like the Dunkers and Old Order Amish of North America. These sects were founded by small numbers of migrants from their much larger congregations in central Europe. They have since remained nearly completely closed to immigration from the surrounding American population. As a result, their blood group gene frequencies are quite different from those in the surrounding populations, both in Europe and in North America."[107] As a result of this genetic isolationism, the Amish possess a uniquely high percentage of those who suffer from Ellis-van Creveld syndrome, a disease that can result in polydactyly (extra fingers and toes).

A similar instance exists among the offspring of a small group of fifteenth-century Ashkenazi (European) Jews whose descendants, as a result of their own insular natures, to this day run a greater risk of acquiring such genetically conceived neurological diseases as Gaucher's, Niemann-Pick, and Tay-Sachs. Moreover, a recent article released by the *Journal of Biosocial Science*, published by Cambridge University Press, speculates that these particular diseases in relation to the Ashkenazi's unique gene frequencies may be associated with an inherent predisposition for higher intelligence, thus postulating that certain genetic strains brought about by genetic isolation might influence specific cognitive traits. This notion is further confirmed by the geneticist L. B. Jorde who wrote, "Many geographic, climatic, and historical factors have contributed to the patterns of human genetic variation seen in the world today. For example, population processes associated with colonization, periods of geographic isolation, socially reinforced endogamy (intermarriage), and natural selection all have affected allele frequencies in certain populations."[108]

With all this in mind, couldn't we surmise that were we to take a sampling of hyper-religious individuals and sequester them on an island that generations later their descendants might be highly religious as well? Consequently, isn't it therefore possible that, as a result of genetic drift, the original pioneer communities of colonial North America brought with them enhanced "religious" genes furnishing their progeny with amplified predispositions toward increased religiosity?

As my brief historical account of the colonies indicates, America's original pioneers were predominantly composed of Europe's zealots, the devout, the steadfast, fanatics, fundamentalists, those who resisted assimilation with the accepted religion of the time, those who defied inquisitions, persecutions, executions, and exile just so they could retain their religious faiths. In the face of having to choose between either cultural assimilation or possible death via imprisonment, execution, or banishment, isn't it safe to presume that only the most devout–the hyper-religious–would choose such a treacherous path?

The colonial immigration to North America represents a unique event in human history in that it is perhaps the greatest mass migration of various peoples from various lands all to one place motivated by one specific social agenda–to preserve their religious faiths. And although the majority of the descendants of these early American pioneers may not have lived quite as insular existences as the Amish, "during the first two centuries of its existence, New England was unusually homogenous in its population characteristics."[109] This homogeneity was reinforced by the fact that for years after the revolution–up until the 1830s–immigration was kept down to less than a few thousand a year, so that "from the time of independence, immigration had accounted for little of the nation's population growth."[110] By 1830, of America's total population of nearly thirteen million, fewer than five hundred thousand were foreign-born.

From the mid-1800s on, American history is comprised of an assortment of ebbs and flows in immigration so that between 1820 and 1992, an additional fifty-seven million immigrants were added to America's total population which by that year was approximately two hundred fifty-five million (realize that this does not mean that nearly two hundred million are pure-stock pioneers, as one must account for the fact that these fifty-seven million immigrants have been procreating along the way so that much of the two hundred fifty-five million constitute their progeny). As a result, to attempt to calculate the genetic interplay between later immigrants and the earliest settlers would be next to impossible.* Even so, with all the

*In delving into the murky waters of population genetics, there are so many variables to consider, so much demographic interplay and mixing of genetic material that it is practically impossible to draw any certain conclusions. In addition, the passing of specific behavioral traits among particular groups is, in itself, an entirely conjectural science. Whether we are discussing the possibility of Jews being inherently smarter or Asians being more inherently predisposed to math or science, though these often represent cultural realities, it is merely speculative–and in some cases dangerous–to draw genetic conclusions. At the same time, however, as we know that certain proclivities can be passed from generation to generation, it should also be recognized that the subject warrants consideration.

blending of America's melting pot of a gene pool, it is still estimated that "the old-stock English Protestants comprise about 45 percent of today's U.S. population,"[111] indicating that a significant degree of homogeneity has been preserved to this day. After all, if after five hundred years of being dispersed around the globe, often intermarrying along the way, the Ashkenazi Jews can retain part of their genetic identity, why can't the same be true for America's pioneers?*

Granted, there have been many additions made to American's gene pool since colonial times, countless numbers of individuals who came to the states with no interest or preoccupation with religion whatsoever. Nevertheless, it could be argued that the seeds of religiosity had, by this time, already been sown and diffused into our nation's cultural landscape as well, perhaps, as its gene pool, enough to make the kind of impact we see reflected in our present-day religious statistics. Today, the United States boasts more religious cults and sects than any other nation on earth with over fifteen hundred primary religious denominations, more than two hundred exclusively Christian television and radio stations, more than three hundred thousand local congregations, and over five hundred thirty thousand total clergy, conspicuously more than any other nation–a testament to American freedom and diversity, and perhaps, to some extent, the nature of our genetic hardwiring.

*Also realize that of the statistics cited in the case for Jewish intelligence (e.g., the disproportionate number of Nobel Prize–winners), these are reflective of the world's secular Jewish community, who are generally much less insular and therefore much more open to intermarriage than their endogamous Orthodox counterparts who are really the ones with more isolated gene pools.

Chapter 18

The Guilt and Morality Functions

"Scientists and humanists should consider together the possibility that the time has come for ethics to be removed temporarily from the philosophers and biologized."[112]

–E. O. Wilson

"There are no moral phenomenon but only a moral interpretation of phenomena."

–Nietzsche

Just as individuals from every culture have possessed the capacity to experience feelings of sadness, individuals from every culture have possessed a capacity to experience feelings of what we call guilt–a remorseful awareness of having done something wrong. This would suggest that the experience of guilt represents yet another genetically inherited characteristic of our species. We can therefore presume that there must exist some neurophysiological mechanism from which this experience is generated, thereby also intimating that we must possess what we could call "guilt" genes that prompt our emerging brains to develop those neural connections that will come to constitute such a "guilt" mechanism in us. But what is the origin of such a peculiar sentiment in us? What is its function? Furthermore, in what way might this sentiment be related to our spiritual functions?

In order to understand the nature of guilt, we must first chart its evolutionary origins. During the time of the emergence of organic matter, the majority of Earth's life forms lived independently from one another as opposed to in groups. This was primarily due to the fact that during those earliest times, all life reproduced asexually and, consequently, had no real need to congregate. In asexual reproduction one genderless, single-celled organism spawns another by forging an exact duplicate of itself. Due to the nature of this reproductive strategy, there was never any need for any two organisms from the same species to interact.

As life continued to evolve, however, two distinct sexes emerged. Among these new sexually reproducing organisms, it now took two members of the same species–one of each gender–to merge their genes in order to procreate. This new reproductive strategy served to an organism's advantage in that it promoted greater diversity among offspring. Greater diversity meant a greater chance for more advantageous adaptations to emerge. The more advantageous adaptations that emerged, the more a species was likely to survive.

Even with the advent of sexual reproduction, the majority of species were still non-social, meaning each individual organism still lived a predominantly solitary existence. The difference now was that the two sexes had to meet at least once in a lifetime in order to procreate. Such gatherings often occurred during a species' mating season, in which the two sexes met, usually for the first and only time, merely to copulate and then go their separate ways. Moreover, among such species, once the mother laid her eggs, she usually abandoned them, never to behold her own progeny.

As time went on and life continued to diversify, an evolutionary trend began to occur in which individual organisms started to live among one another in groups. Within a group, each individual organism was more secure than if it lived on its own. Within a group, not only could individuals better defend themselves against predators, but they could more effectively hunt and forage. Because of the strength and stability that came with this social adaptation, the group dynamic became the "favored" evolutionary trend, particularly among vertebrates and most particularly among mammals.

With all the advantages that came with this new group dynamic, there were some disadvantages as well.* In order to put some perspective on the disadvantages of the group dynamic, we need look at this adaptation's origins. Before the emergence of the group dynamic, individual organisms lived primarily by and for themselves. Because these earliest life forms lived exclusively solitary existences, they did so without regard for any other member of their species. Consequently, all behavior was governed by an animal's self-serving instincts. It was a strictly planarian-eat-planarian world.

As organisms evolved to coexist among one another in groups, these selfish instincts no longer served to an animal's advantage. Obviously, if every creature within a group only struggled for its own preservation without any regard for any other individual within its community, it would be impossible for such a group to survive. Now that life forms were evolving to coexist among one

*There is no such thing as a perfect trait. For every adaptation, as advantageous as it might be, there is always some drawback. For example, though the sickle cell was selected in humans for its ability to help us resist malaria, its emergence constituted its own threat. In this way, evolution works as a seemingly haphazard process of trials and errors. As new variations emerge with each individual organism, some are to the individual's advantage, some to its disadvantage, while almost all are a little of both. In essence, every trait we possess comes with its share of pros and cons. In accordance with the essential physical laws of nature (e.g., the laws of thermodynamics), we could say that any given variation to emerge renders an organism either more or less energy-efficient. Whereas those variations that happen to be more energy-efficient are most likely to endure, those which are least so are most likely to succumb to the forces of extinction.

another in tightly knit groups, newer adaptations had to emerge by which a species could balance the needs of the individual with the needs of the community. In other words, organisms had to evolve a capacity to apportion their own needs so that they could serve themselves while simultaneously serving the needs of their group. Strictly selfish behaviors suddenly represented a threat to the group, which, in turn, represented a threat to every individual within that group. Though each individual added to its group's strength and therefore served to its advantage, because each individual also possessed its own set of self-serving instincts, each member simultaneously represented a potential threat.

This was not the only drawback to arise with the emergence of the group dynamic. Now that individual organisms lived in such close proximity to one another, there was an increased likelihood of transmitting contagious diseases. Among the less social species, one single organism infected with a transmittable disease was much more likely to die on its own without infecting another of its own kind. Since these social organisms lived in such close contact with one another, now when an individual was infected with a transmittable disease, it was much more likely to infect the entire community.

A third problem of the group dynamic was that it represented a potential threat to a species' gene pool. Since the group worked to protect all of its members, now even the weakest members of the species were more likely to survive. On its own, a weak, sickly, or handicapped organism is less likely to survive. Among the group dynamic, however, even the weakest members are at least partially safeguarded by the group from any external threat. Consequently, among the social orders, it became more likely that a weaker individual might live long enough to reproduce and therefore to pass its "inferior" genes on to future generations, thus negatively affecting the group's as well as the entire species' gene pool.

Suppose, for example, an organism from a non-social species happened to be born with a bad leg or inferior vision. In such cases, not only would that individual find it difficult to hunt or forage, but it would have an equally difficult time safeguarding itself against predators. Among the group, however, this same physically handicapped individual would have a much better chance of surviving since it would be sheltered by the group. Therefore, though the group dynamic represents a highly advantageous adaptation, it at the same time threatens to compromise a species' gene pool.

Among those organisms that live solitary existences, the weakest are more vulnerable and therefore less likely to survive. In this way, with every passing generation, the weakest members of a species are weeded out (along with their genes) for extinction. As a result of this dynamic, with every passing generation, every species should be that much better suited to meet the demands of its physical environment than the one that came before. It should be stronger, more fit (energy-efficient), and therefore more likely to persevere.

Among the social species, however, this principle no longer applied. Among such species, the rule becomes survival of the fittest as well as the weakest. Among the social species, the law of survival of the fittest–the principle that guides all natural selection, all organic evolution–becomes compromised. As a result, the chances of any such species surviving is compromised as well.

As advantageous as the group dynamic may have been, by safeguarding the weakest members of each social species' gene pool, it threatened to derail the selection process. Among the social animals, rather than a species' gene pool getting stronger with each passing generation, it now remained stagnant. To compensate for these drawbacks, newer adaptations had to emerge among the social organisms.

To circumvent these new obstacles, the social organisms began to develop new mechanisms which enabled them to counter these problems. One such mechanism to emerge took shape in what is called "ostracizing" behaviors. Here, the social species evolved a mechanism that enabled them to distinguish genetically healthy individuals from diseased, handicapped, or generally unhealthy ones.*

Once these social animals had evolved a mechanism by which they could recognize a defect (a disease or disability) in another member of their species, a supplementary mechanism had also emerged that now compelled those same creatures to be repulsed by such physical irregularities. This is manifest in the way that healthy organisms will instinctively shun, avoid, and in some cases even become belligerent toward a weak, diseased, or handicapped member of its species. Such behavior can be witnessed among the young of many mammals who tend to shun, torment, and in some cases even kill the weakest or "runts" of their own litters. Among our own species, which is perhaps the most discriminating of all, ostracizing behaviors are most apparent in children, as they have yet to be sufficiently socialized to behave more sympathetically towards a mentally or physically handicapped individual.

This ostracizing mechanism helped to resolve two of the most essential problems associated with the group dynamic. Being that

---

*There are those who hypothesize that many organisms detect physical health in others of its species through the visual recognition of symmetry in the physical features of that organism. Physical symmetry, it has been suggested, correlates to fitness and therefore becomes the mechanism by which many animals discern a healthy individual from a diseased or handicapped one. For example, an animal's limp or hunchback, both which would compromise an animal's symmetry, represent visual indicators of a genetic defect. Among our own species, this same mechanism might be responsible for determining our aesthetic sensibilities by which we call some individuals "beautiful" as compared to those we call "ugly." To confirm this notion, Victor Johnston, a psychologist at New Mexico State University, conducted a study in which he used electrodes to see what happens to the brain's electrophysiology when we look at different faces. What Johnston found was that when people look at a symmetrical female face as opposed to a less symmetrical one, the brain becomes much more excited. Apparently, the visual detection of symmetrical features, what we otherwise refer to as beauty, seems to have neurophysiological consequences. Consequently, physical attraction must be neurochemical in nature. We can therefore say we are drawn to beauty like a drug. This might help to explain, for instance, why billboard, magazine, and TV ads from almost every culture are inundated with images of beautiful women who are used to draw us in as a means to help sell their products. Apparently, just as is true with love, morality, or God, it appears that beauty, too, is a relative concept determined by our "wiring."

many diseases reveal themselves by affecting our physical appearances (scabs, open sores, infections, distressed complexion, weakened constitution, bloodshot eyes, etc.), social animals now ostracized the sickly, thus helping to deter the spread of transmittable diseases. Second, the ostracizing reflex prompted social animals to cast out those members of their communities with substandard genes, fortifying the group's as well as the entire species' gene pool.

Even with these two threats resolved, there still existed that internal threat to the group generated by those destructive, yet necessary, selfish instincts inherent in each individual within the group. How was nature to balance the conflicting needs of individual self-preservation with the need to preserve the group? Obviously, no organism could survive if it lost all of its self-serving instincts and lived exclusively for the welfare of others. At the same time, no group could survive if each member was exclusively self-serving and completely inconsiderate of the needs of others within its community. For this reason, nature had to select a new mechanism that would balance these two essential yet conflicting needs.

In pre-human social orders, the threat posed to the group by individual selfish behavior was held in check by an evolutionary strategy known as the hierarchy system. In hierarchy systems, each member of the group engages one another in a series of physical contests (this doesn't necessitate contact but can be resolved merely through physical gesturing and posturing) until each individual's rank in the hierarchy is determined. Whichever individual proves itself strongest of all will dominate the others as their leader. This dominant individual (often referred to as the alpha male or female) will be first in line to eat when food is procured. More significantly, he or she will also have first choice in the selection of a mate. This will ensure that the fittest male's genes will be coupled with the fittest female's, ensuring the production of the fittest offspring.

Despite the fact that the group was comprised of individuals generally driven by more selfish instincts, the hierarchy system maintained stability and order within groups. In such a dynamic, though a member of the group might at times be tempted to act on his or her more selfish impulses, such instincts are held in check by the structure of the hierarchy. Should an individual, for instance, try to take more than its fair share of a kill, that individual will inevitably be challenged by one of its superiors. Should this "greedy" individual desire to dispute its rank, it can at any time challenge another member of its group to a physical contest. If the challenger prevails, its position in the group is elevated. If it loses, it will either maintain its old rank or, in some cases, it might even be shunned or chastised by its community for trying to usurp a superior and disrupt the group order. Amid the hierarchy system, the group dynamic was maintained by the simple law of domination by the fittest. At no point, for example, could a weaker member claim superiority without eventually being challenged and forced back into submission. In this way, physical strength settled all scores and helped to maintain a harmonious order among the pre-human social species.

With the advent of humans, however, all this changed. Humans, in a sense, represent the end of the physical hierarchy system. Unlike any other species, because of our cerebral capacities, every individual possesses the power to subjugate or kill any other. Before humans, if a weaker member within a group challenged a superior, he or she would be defeated based on pure physical strength. With the emergence of human intelligence, however, even the physically weakest member of a community possesses the capacity to kill, and, consequently, to displace any other. Among human societies, even the physically weakest member of a community can, for example, should he or she be so inclined, pick up a heavy object and bludgeon the physically strongest member of its community to death. With our enhanced capacity to devise and construct tools, the lines of the hierarchy became irrevocably blurred. In light of our intelligence, power took on a whole new meaning. No longer could a human society rely on raw, physical strength to maintain social stability. Instead, some newer device was now needed if the group, not to mention the entire species, was to survive. It was at this point in our evolution that a moral function emerged.

Just as all cultures display a distinct set of what we could classify as "spiritual" behaviors, all cultures display a distinct set of what we could classify as "moral" behaviors. Moral behavior can be characterized as that tendency in our species (and only our species) to categorize every action as being either productive or destructive to the group's welfare. Those acts perceived as productive to the group are cross-culturally classified as "good," while those acts we perceive as harmful to the group are classified as "bad." This propensity to discern "good" from "bad" behaviors is made evident by the fact that every culture has compiled lists of rules and regulations (laws) in which "good" acts are encouraged and destructive or "bad" acts are discouraged. Just as our biological ancestors ostracized those individuals who represented a threat to the group, we do the same, only in a more sophisticated way.

Though our species may possess some very strong communal instincts, we are still driven, to a significant extent, by our more selfish and destructive impulses. Consequently, it became necessary for our species to evolve a moral function. Just as our ancestors could distinguish a physically healthy and fit individual from a diseased or handicapped one, because our species is so much more behaviorally complex, it became necessary that we develop a capacity to distinguish healthy behaviors from unhealthy ones. Again, those behaviors we perceive as being advantageous to the group, we define as "good," whereas those we perceive as harmful, we define as "bad."

By implementing our language functions, humans possessed the capacity to compile verbal and, eventually, written lists of those behaviors they perceived as being potentially harmful to the group. Once these rules became codified, the group was compelled to ostracize or punish any individual who transgressed one of its "laws." To enforce these laws, we developed an instinct to punish those who broke them. In essence, humans had evolved a penal function to complement our moral one. This penal function represents that impulse in us to systematically ostracize and/or punish those who transgress our society's laws. For the majority of our species, fear of such punishment inhibits individuals from acting on his or her more selfish instincts. Once we evolved this instinct to enforce our laws, group order could survive despite our more selfish impulses. I imagine that if such a function hadn't emerged in us, the group dynamic, not to mention our entire species, would have most probably succumbed to the forces of anarchy and with it extinction.

Though our entire species possesses the same language centers in the brain, every culture, based on its own particular historical and environmental circumstance, has developed its own specific language. Though each language may be unique, each contains certain universal characteristics. Likewise, though our entire species possesses the same spiritual/religious impulse, every culture, based on its own particular circumstance, has cultivated its own unique religion. Again, as unique as each religion might be, they all possess distinct similarities. Analogously, though our species may possess the same moral function, every culture, based on its unique circumstance, has developed its own moral code, though beneath the seeming differences all have distinct similarities. For instance, incest and murder represent universally proscribed behaviors, otherwise known as taboos. The reason such universal taboos exist is because we, as a species, are neurophysiologically hardwired to be repelled by such acts. It is necessary we be "wired" this way as such acts constitute an obvious threat to the group dynamic.

The first clue to reveal that we might be hardwired for moral behavior can be traced to the bizarre case of Phineas Gage, a railroad worker who, in 1848, was the victim of a dynamiting accident that drove an iron rod straight through his skull. Though Gage survived the accident without suffering any noticeable damage to his intellect, his personality had been radically changed. Prior to his accident, Gage was known as an honest, upstanding family man and a diligent worker. Weeks after his accident, however, he became an irresponsible and unethical drifter, prone to lying, cheating, and stealing. Later studies revealed that the spike had gone through Gage's prefrontal cortex, indicating that this part of the brain might play a crucial role in moral and social reasoning, thus paving the way for a neurobiological interpretation of moral consciousness.

Recent studies conducted by Antonio Damasio of the University of Iowa offer new evidence supporting this notion:

> Damasio and colleagues found two subjects who suffered damage to their prefrontal cortices before the age of sixteen months. Both children seemed to recover. But as they aged, the two began to behave aberrantly–stealing, lying, verbally and physically abusing other people, poorly parenting their out-of-wedlock children, showing a distinct lack of remorse, and failing to plan for their futures.[113]

Moreover, it appeared that there was no obvious environmental explanation for the youths' behavior as both had been raised in stable, middle-income families and had well-adjusted siblings. Based on his research, Damasio concluded:

> Early dysfunction in certain sectors of prefrontal cortex seems to cause abnormal development of social and moral behavior, independently of social and psychological factors, which do not seem to have played a role in the condition of our subjects.[114]

To provide further support of Dr. Damasio's findings, Drs. Ricardo de Oliveira-Souza and Jorge Moll of the Neurology and Neuroimaging Group, LABS and Hospitais D'or, Rio de Janeiro, Brazil, used magnetic resonance imaging (MRI) to reveal those parts of the brain that become activated when a person contemplates ethical concerns. This was accomplished when ten subjects, a mix of men and women, ranging in age from twenty-four to forty-three, were asked to make a series of moral judgments while inside an MRI scanner.

> On headphones, the study participants listened to a series of statements, such as "we break the law if necessary," "everyone has the right to live," and "let's fight for peace." In each case, the subjects were asked to silently judge if each sentence was "right" or "wrong." The participants also listened to sentences with no moral content, such as "stones are made of water" or "walking is good for health," and judged these in a similar fashion.[115]

Results from brain scans taken as the subjects were in the midst of contemplating such ethical problems showed that the moral decision making process was associated with the activation of a region within the brain's frontal poles known as the Brodmann area 10 or the mid-dorsolateral prefrontal cortex. In accordance with Dr. Damasio's results, the researchers who conducted these MRI experiments also found that "people who injure this area of the brain may exhibit severe antisocial activity."[116]

Furthermore, it seems that we possess a proclivity to project our spiritual conceptions onto our moral ones. For instance, behaviors that are looked upon as "good" are, in a spiritual context, perceived as "pious," "virtuous," or "holy" and are seen as being looked upon favorably by our gods. At the same time, we seem to be equally inclined to perceive destructive or "bad" acts as condemned by our gods. Those actions we might label as "bad" are, in a spiritual context, cross-culturally referred to as what we call "evil," a concept for which every known culture has possessed a symbol or word. To support this notion, every culture has maintained a belief in "evil" powers or entities (e.g., demons) whose purpose is to tempt the fate of our immortal souls as well as to inflict harm and suffering on us. In addition, almost every world culture has conceived of a place where the souls of those who commit "evil" deeds are condemned to suffer eternal damnation.

Hell, Niflheim, Tartarus, Gehenna, Jahannan, Bhumis, Karmavacara, and Hades are examples of places that different world cultures believe "evil" souls are sent after death.

On the other hand, the souls of the "good" are cross-culturally perceived as being rewarded by our gods. Whether it be Heaven, Nirvana, the Happy Hunting Grounds, Valhalla, or the Elysian Fields, almost every world culture has believed in a place where "good" souls are rewarded in the afterlife. All of this suggests that moral consciousness must be integrally linked with spiritual consciousness. In light of this, moral consciousness, just like spiritual consciousness, must be viewed as nothing other than the manifestation of another genetically inherited impulse, another inherent component of human cognition. Consequently, such notions as "good" and "evil" must be viewed, like all physiologically generated perceptions, as subjective conceptions relative to the particular manner in which our species happens to be "wired" to perceive and interpret reality, and not something founded in some absolute or transcendental truth.

Even with the emergence of a moral and penal impulse, our species' selfish instincts still tempted us to defy our society's laws. It was here that "nature" selected two more mechanisms by which we could balance our selfish impulses with the needs of the group.

The first of these new adaptive impulses to emerge in us was an altruistic drive. In order to balance our selfish impulses, nature installed a device in our species that countered our instinct to serve ourselves with one that compelled us to serve others within our community. With the addition of an altruistic impulse, humans were now compelled to serve others with nearly the same determination with which they were compelled to serve themselves.

As with any trait, each individual possesses this altruistic impulse in varying degrees. Though the average person may possess an average proclivity to engage in altruistic behaviors, there exist those individuals who possess either a diminished or an enhanced propensity for this impulse. On one extreme, every culture contains a certain percentage of individuals who are "wired" with an underdeveloped altruistic impulse, those who are much more motivated by their self-serving instincts. Such individuals might be represented within our societies by its selfish and greedy, its robber barons, exploiters, misers, and thieves, people with little or no regard for others within their community and who are only capable of looking out for their own self-interests, those of whom we might say possess no social conscience. For such people, the desire to give or assist others does not play a significant role in their conscious experience.

On the other extreme, each culture contains a small percentage of individuals who possess an overdeveloped altruistic drive and who possess a very strong impulse to give. Such individuals are more likely to be found playing the role of social reformers, philanthropists, missionaries, and aid and welfare workers, as a few examples. Such individuals are, for better or worse, often compelled to be more concerned with the welfare of others than they are for their own selves.

The second trait selected to help us temper our more selfish instincts, I refer to as the guilt function. As mentioned earlier, individuals from every culture have shown a capacity to experience feelings of guilt, suggesting that a "guilt" mechanism must have emerged in our species to complement our moral and altruistic drives. Whereas our moral and penal functions provide us with a means to discern and then shun and/or punish others who act on their more selfish instincts, the guilt function provides us with a mechanism that compels us to shun and/or punish our own selves for committing the same selfish acts we find reprehensible in others. Just as our nervous systems prompt us to retreat from such potential hazards as fire, this sentiment of guilt provides us with a mechanism that prompts us to instinctually retreat from committing such potentially hazardous social acts as murder, incest, and stealing. Though many selfish acts might momentarily serve to our advantage, they represent a threat to the group dynamic, which, because we are all a part of the same group, ultimately represents a threat to ourselves. Ironically, it is in our self-interest not to be overly selfish.

With the advent of guilt, our moral functions had become internalized in such a way that we were now "wired" to be just as repulsed by our own selfish propensities as we were by those of others. By constantly carrying around these internalized self-critical impulses, each individual was forced to become ever-vigilant over his own selfish instincts.

Just as with all other traits, each individual is predisposed to experience guilt in varying degrees. Though the average person of any population will most likely possess an average capacity to experience guilt, each culture possesses a smaller percentage of individuals who represent the extremes of this sentiment. On the one hand, there exist those born with an underdeveloped guilt function, those whom, no matter how much society may try to change them, are incapable of experiencing feelings of remorse. These are represented by a society's socio/psychopaths–individuals who have a clear grasp of reality but are capable of committing selfish acts without experiencing remorse, those who we might say possess no social or moral conscience. Because such individuals are not compelled to contain their selfish impulses, they often constitute a society's criminal element.

According to Nicholas Regush, author of *The Breaking Point: Understanding Your Potential for Violence*, statistical research has revealed that a cross-section of every culture demonstrates psychopathic tendencies, revealing that the origins of this psychosocial disorder may stem from the workings of the brain. "A common estimate is that about 1 percent of the general population is 'psychopathic,' as well as perhaps as much as 20 percent of the prison population."[117]

To support a neurophysiological explanation of psychopathic behavior, the psychologist Robert Hare at the University of British Columbia reported:

> In psychopaths there appears to be less than normal use of brain regions that integrate emotions to memory with other brain functions. The researchers reached their conclusion by comparing

> brain waves of subjects deemed to be psychopaths with the brain waves of so-called normals. The data were gathered during the performance of a language test that required responses to neutral and emotionally laden words. Research elsewhere with brain scans has since shown that when psychopaths responded to the emotional words parts of their brain, such as those regulating emotions (the amygdala) and long-term planning (a region of the frontal cortex), remained inactive; these brain regions in normals were active when they responded to the same words.[118]

Just as there exist those who are incapable of feeling guilt, on the other extreme of the guilt bell curve, every culture maintains a cross-section of individuals who possess an overactive guilt function. Such individuals are plagued with excessive feelings of guilt, regardless of whether or not they have done anything wrong. These overly self-critical or guilt-filled individuals feel a constant need to criticize, condemn, and punish themselves. In the words of Karen Horney, such an individual "insists on his guilt and vigorously resists every attempt to be exonerated." Those who suffer from this particular cognitive dysfunction are often represented by society's penitents and ascetics, those who have a tendency to be self-flagellating as well as self-deprecating and who tend to refrain from indulging themselves as they feel a constant need to punish and deprive themselves.

More evidence to support a genetic interpretation of guilty behavior is found in the fact that delusions of guilt–hallucinations involving having done something wrong or "sinful"–represent a common symptom of schizophrenia. That this particular delusion emerges as a cross-cultural symptom of what we know to be a neurophysiologically based disorder suggests that the experience of guilt is neurophysiological in nature, an inherent part of human cognition.

So what relationship might our guilt function have with our spiritual one? Generally speaking, when we commit a wrongful act, our guilt is directed toward the victim of our misdeed. At the same time, however, humans have a distinct propensity to also feel guilty for their wrongdoings before their gods. This is made evident by the fact that every culture has conceived of the notion of "sin." When we transgress the laws of our community, we call it a crime. When we transgress what we perceive to be the laws of our gods, we regard it as a sin. The fact that every culture has possessed a word to express this concept suggests that feelings of guilt–which increase our anxiety levels–have a tendency to incite religious consciousness.

To further support this notion that our guilt function is integrally linked with our spiritual/religious ones, all cultures have maintained rites through which we seek to repent or atone for our sins. Such penitent behaviors are clearly related to the sentiment of guilt.

When the average person commits a wrongful act, it seems to evoke a great deal of anxiety. Much of this anxiety can be attributed to the fear of social as well as divine retribution. Moreover, anxieties evoked by guilt have a tendency to stimulate spiritual/religious consciousness, thus having the effect of turning men to God. This may help to explain, for instance, why prisons often contain such an abundance of religious converts.

> Moral anxiety based on guilt and guilt feelings activates religious concerns...In fact, the existence of morality is, to many people, impossible without established religion and belief in God.[119]

When we contemplate or commit an antisocial act or "sin" such as murder, it evokes an unpleasant sensation meant to deter us from acting on such impulses. Because these sensations are mysteriously evoked from within, we tend to interpret them as evidence that we are being punished or haunted by our gods. Moreover, as a result of the innate nature of these feelings, we tend to interpret such principles as "thou shalt not steal or kill" as self-evident truths which have been laid down to us by some divine or transcendental authority.

It is for this same reason that many believe that being moral is contingent on believing in an established god or religion. It is also for this same reason that atheists are often stigmatized as being inherently immoral, something I contend is nothing more than an unfounded bias. As Einstein expressed this same sentiment, "A man's ethical behavior should be based effectually on sympathy, education, and social ties; no religious basis is necessary. Man would indeed be in a poor way if he had to be restrained by fear of punishment and hope of reward after death."

Though an atheist might not be physiologically hardwired to possess strong religious or spiritual inclinations, his moral centers might be more developed than an overtly religious and/or spiritual person. Again, we are talking about three distinct intelligences, three types of "wiring" (moral, spiritual, and religious), three modes of consciousness that can be as unique to one another as our faculties for language, music, or math. It's therefore no more likely that an atheist should be immoral or sociopathic than someone who believes in God. Consequently, religion and morality should not be viewed as any more synonymous than should atheism and immorality. To counter this stigma, some atheists refer to themselves as "secular humanists" to define their sense of moral and social responsibility.

Chapter 19

The Logic of God: A New "Spiritual" Paradigm

"We are what we think. All that we are arises with our thoughts. With our thoughts we make the world."

—Buddha

"Projection makes perception. The world you see is what you gave it, nothing more than that. It is the witness to your state of mind, the outward picture of an inward condition. As a man thinketh, so does he perceive. Therefore, seek not to change the world, but choose to change your mind about the world."

—Anonymous

"The real voyage of discovery consists not in seeking new landscapes, but in having new eyes."

—Marcel Proust

"An evolution of consciousness is the central evolution of terrestrial existence...a change of consciousness is the major fact of the next evolutionary transformation."[120]

–S. AUROBINDO

So, what if Kant was right? What if all of our conceptions of reality are really nothing more than the products of internally generated cognitions, sensations, perceptions, "the outward picture of an inward condition"? In such a light, we must accept that all we interpret as being "real" or "true" is subjective, relative to the manner in which our species is hardwired to perceive the world.

Because each species processes information differently, each species consequently interprets reality from its own unique perspective. As all of our perspectives are relative, no species, nor any individual within a species, can ever claim that its interpretation of reality constitutes any absolute truth. As Kant expressed it, we can never possess absolute knowledge of "things in themselves," but only relative knowledge of "things as we perceive them." Just as flies possess fly knowledge, humans possess human knowledge. And just as flies possess fly "truths," humans possess human "truths," neither being any more genuine or "real," just different. We must therefore accept that such notions as absolute truth are incomprehensible ideals. Instead, we are forever bound to our relative human perspectives which are framed by the way our brains process information. Consequently, to understand the nature of human reality, we first need to understand the underlying nature of how our brains work.

The human brain consists of an interactive network of separate regions, each that processes information in a unique way. These are our cognitive functions. We have a language function (based in the Wernicke's area, Broca's area, and angular gyrus), an anxiety function (based in the amygdala), a morality function (based in the mid-dorsolateral prefrontal cortex), and the list goes on. Essentially, for every sensation, perception, cognition, or behavior our species cross-culturally experiences or engages in, there is some specific region in the brain responsible for generating that specific function. Consequently, in order to better understand the nature of how our brains process information, we need to learn the nature of each of those individual cognitive functions from which we derive the sum of our conscious experience. It is the role of each of these distinct cognitive functions to process a multitude of data, each in its own particular way. Only after all of this separately processed data is integrated are we provided with a comprehensible picture of that which we refer to as reality.

So what if we were to apply this same precept to human spirituality? What if spirituality represents the manifestation of one of these cognitive functions, one of our brain's inherent modes of cognitive processing? As all cultures perceive a spiritual realm, isn't it possible that spirituality may represent one of the ways our species is "wired" to process information and consequently to interpret reality? If so, this would imply that our cross-cultural "spiritual" beliefs in such concepts as a god, a soul, and an afterlife constitute nothing more than manifestations of the way our species happens to process information and therefore interpret reality. In such a case, God would no longer represent any absolute being but rather a cognitively generated, subjective, human conception–not a divine but an organic phenomenon. In essence, God, as we've thus far interpreted him–as a real and absolute entity–is, as Nietzsche suggested, dead. No longer an absolute reality, God is reduced to just another one of our species' relative perceptions, the manifestation of an evolutionary adaptation–a coping mechanism–instilled in us to help us endure life's hardships as well as the otherwise debilitating awareness that we must die.

I realize that it may be difficult for many people to accept such a reductionistic/evolutionary/organic/cognitive/rational, that is, scientific, interpretation of God. Because the majority of our species is hardwired to perceive a spiritual realm, it may literally be impossible for many to even grasp this concept as it may conflict with their inherent perception of reality. Subsequently, trying to convince someone who is hardwired to believe in a spiritual reality that no such thing exists may be as futile as trying to convince a schizophrenic that the voices he hears are coming from within his own head as opposed to from some external reality. This is not to suggest that our spiritual perceptions represent a physical dysfunction, as is true of schizophrenia. On the contrary, spiritual consciousness represents a normal part of the human cognitive experience.

But what if we could somehow get the schizophrenic to recognize that his hallucinations are nothing more than the products of internally generated misperceptions? What if we could teach him to reason through his delusions? Similarly, what if our entire species could be taught to reason through our delusional beliefs in the supernatural? What if we could come to recognize that such beliefs aren't representative of any actual transcendental reality but are, instead, the manifestations of internally generated misperceptions: God as a cognitive phantom. What if we could recognize that spiritual consciousness exists as the consequence of a neurophysiological reflex? Just as planarians reflexively turn towards light, humankind reflexively turns towards imaginary powers.

Imagine an android that is programmed to believe that it's human. In order to make the android believe such a thing, the manufacturer installed a computer chip into its circuitry which instilled it with fictitious memories of a fabricated past (similar to the plot of the film *Blade Runner*). Now imagine that the android were to suddenly become aware of its true nature (also similar to *Blade Runner*). Suddenly, it realizes not only that it's an android, but that its memories are nothing more than the effects of a computer chip that compels it to perceive a delusional past. Now that the android has become cognizant of its true nature, it would be free to explore the possibilities of a whole new paradigm. No longer bound to the false reality with which it was preprogrammed, the android would now be able to redefine its own destiny, able to explore new possibilities in accordance with its "truer" nature.

Analogously, imagine humans were to suddenly become cognizant of the fact that they've been preprogrammed by the forces of nature to perceive a spiritual reality, one that is just as fabricated as our android's fictitious past. Just as the android had been constructed with computer chips that frame its illusory perceptions, humans are analogously constructed with a neural network that frames ours in a similar way. What if in the same manner that our android recognized its memories existed not as the recollection of actual past experiences but rather as the consequence of a computer program installed into its circuitry, we came to recognize that spiritual consciousness exists not as the effect of any actual transcendental reality but rather as the consequence of an organic program–a reflex–installed into our species' neural circuitry? Perhaps if we learned to regard spirituality in such a way, we too could devise a whole new paradigm for ourselves, one through which we could redefine our own destinies based on our "truer" natures. Rather than having to be stuck in the same delusional consciousness nature forged for us, we could use this newfound self-knowledge to strive for a healthier and more productive vision of ourselves.

As another metaphor, imagine we are looking into a mirror that can offer us a pure reflection of ourselves. Now imagine that placed between us and this pure reflection is a series of invisible lenses, ones that will distort our otherwise unadulterated view in some way. Because we are ignorant that these lenses exist, we have no way of knowing that our self-perceptions have been distorted. Though we may believe that our view represents a perfect reflection of ourselves, we are actually misinformed. Not until we become aware that these lenses exist, until we learn to look past them, to push them aside, will we be afforded a true reflection of ourselves.

I believe that human spirituality represents such a lens, one that distorts our view of reality by making us perceive a spiritual element when no such thing exists. But what if we were to become aware that such a lens does exist? What if we were to choose to push it aside, clearing our perspectives of all such "spiritual" distortions, affording ourselves a much clearer, less obstructed view of reality? Sure, it might be somewhat uncomfortable at first, even distressing, to have to readjust our perceptions of ourselves in such a fundamental way. But wouldn't we prefer to possess a more perfect view of reality than a distorted one? Shouldn't we want truth over deception?

Spiritual consciousness constitutes "nature's white lie," a coping mechanism selected into our species to help alleviate the debilitating anxiety caused by our unique awareness of death. But is it even possible that nature would program a species with an inherent misperception, a built-in lie? Truth, lies, reality. These are human conceptions that have no bearing on the manner in which nature framed us. The process of natural selection has no regard for such lofty human contrivances as "real" or "true." Nature's only impetus is to create a more survivable organism, one that can more effectively pass its genetic material onto future generations–this and nothing more. As *The Selfish Gene*'s author Richard Dawkins expressed it, "We, and all other animals, are machines created by our genes. We are survival machines, robot vehicles blindly programmed to preserve the molecules known as genes."

So as terrifying as the prospect of inevitable and irrevocable death might be, if such an organic theory of spirit and God happens to be correct, isn't it in our best interests to embrace it? What if all the various existing religious paradigms are wrong? Is anything gained by living in conscious denial of the truth? Maybe so. Perhaps if we were to strip the average person of his or her religious faith, we'd end up with a great deal of distress and consequent discord on our hands. Maybe without the benefit of this distorting lens in place, we'd lose our survivability. Therefore, before we consider doing away with all of our old paradigms, it would seem we should really weigh the pros and cons of the situation by asking ourselves: What, if anything, is to be gained by embracing a scientific interpretation of spirituality and God?

Chapter 20

What, If Anything, Is to Be Gained from a Scientific Interpretation of Human Spirituality and God?

"Religion is the source of all imaginable follies and disturbances. It is the parent of fanaticism and civil discord. It is the enemy of mankind."

—Voltaire

"Science is the great antidote to the poison of superstition. An ailing world would do well to reach for the right bottle in the medicine cabinet."

—Adam Smith

"A science that comes to terms with the spiritual nature of mankind may well outstrip the technological science of the immediate past in its contribution to human welfare."

–DR. BENJAMIN SADOCK

"We either come to terms with our unconscious drives and instincts–with life and death–or else we surely die."

–NORMAN O. BROWN

Suppose, for the moment, that what I'm suggesting is ludicrous, the scribblings of a frustrated atheist. Suppose there really is a spiritual realm, a creator, a soul, and an afterlife. Suppose that the essence of consciousness really is a soul that will exist for all eternity. Should this be the case, humankind is free from the threat of death. If we truly are immortal, these bodies we presently inhabit constitute nothing more than superficial skins, which, once shed, will be replaced by another or perhaps, better yet, not replaced by anything at all–spirits set loose to explore the cosmos eternally free from the burden of any restrictive physical reality. Regardless of what particular state eternal life might bring, as long as God exists, as long as there is some supreme transcendental force that has endowed us with an immortal soul, humankind is saved.

Presuming then that God does exist, what harm can there be in merely considering the possibility that He does not? If God exists, what's there to lose in pondering His potential nonexistence? If nothing else, why not simply indulge ourselves in a little mental exploration while we pass some of our endless time?

So let's presume for the moment that God does not exist. Let's presume that all notions of God and spirit are merely neurophysiologically generated delusions/cognitive phantoms/confabulations installed within our brain. If so, what might this mean for us as individuals as well as a species? What are the implications of existing in a spiritless and godless universe? Without God, how are we to gauge our conduct? Where are we to find purpose or meaning in our lives? Without God, is all necessarily lost? Are we truly that defeated and hopeless, or is it possible to find meaning and purpose in other ways? Is it possible we might even be able to use this newfound understanding to improve our existences? In essence, what, if anything, might be gained from a scientific interpretation of human spirituality and God?

To answer this, we must first ask: What is it we want from life? What would we want to gain? Moreover, is there any one thing we might all agree upon? Does such a universal goal exist? And if one does, is its fulfillment contingent on God's existence?

So, is there any one thing that every member of our species universally wants out of life? At the recommendation of one of the greatest thinkers in history, I'm going to suggest that such a universal goal does exist. As Aristotle suggested over two thousand years ago, before all else, all humans mutually strive to achieve the greatest amount of happiness out of life. This, he postulated, constituted humankind's summum bonum–its greatest good. According to Aristotle, every action we take is done with the hope that it will bring us greater happiness (or, in accord with a more Buddhist conception, at least minimize our pain and suffering). This, I agree, represents the universal end of all human action. Moreover, it seems this principle should still hold true regardless of whether or not a god exists. After all, under what conditions would human beings ever seek to be less happy, or, inversely, to suffer more pain and hardship? We can therefore say that whether God exists or not, our ultimate goal is still the same. Consequently, without a god, all is not necessarily lost.

Presuming that maximizing happiness/minimizing suffering represents the desired end of all human action, how are we to reach this goal, most particularly, in a potentially godless universe? Just as the procurement of happiness might represent the universal end of all action, is there a universal means by which we might achieve this goal? Since this brief stay here on Earth might represent our one and only shot at existence, it would seem essential that we be able to answer this question within our lifetime.

In seeking a universal key to happiness, I am again drawn to one of the great ancients. As much as they may have disagreed with one another, practically all of the world's recognized philosophers concur in that the key to happiness lies in the acquisition of knowledge (after all, the word philosophy itself means "love of knowledge"). And of all the various forms of knowledge, the greatest, we are told, lies in self-knowledge. Before all else, said Socrates, "Gnothi seauton"–know thyself.

It is only because our species possesses this unique cognitive capacity for self-awareness that humans can even aspire to acquire self-knowledge, that is, to further know themselves. No other species possesses this ability. Consequently, no other creature can recognize its own shortcomings. Because we can recognize our flaws and weaknesses, humans possess the unique ability to modify themselves in such a way that they can turn a shortcoming into a strength. For instance, should we decide that our inability to fly is a deficit, we can build ourselves wings. Should we feel that we're not fast enough, we can invent the wheel and motor engine, enabling us to move faster than any other creature on Earth. As a result of this capacity, humans can make themselves better suited to their environments, and the better suited we are to our environments, the more apt we are to survive. The more apt we are to survive, the more secure we feel in the world. The more secure we feel, the less anxious we are. The less anxious we are, the happier we will be. In this way, humans possess the unique capacity to modify ourselves in such a way that we can alter ourselves in ways that will make us happier.

As another example of how we can modify ourselves physically, as alluded earlier, should another ice age occur, rather than having to wait millions of years for nature to select a thicker coat of hair for us, we will be able to sew ourselves one in a few hours' time. On a more individual level, a man recognizes that he is physically weaker than his peers. To compensate for this physical shortcoming, he can do any number of things from lifting weights to increase his strength to developing some other capacity, such as his intellect, as a means to more effectively compete with his peers. The more effectively a person can compete with his peers, the more secure one feels; the more secure, the happier.

As an example of how we can modify ourselves not physically, per se, but behaviorally, let's take a man who finds himself lonely in the world and consequently unhappy. After contemplating his circumstance, he realizes that much of his loneliness exists as the result of his selfish tendencies, something that has driven away most of his family and friends. In recognizing that his selfish ways represent the chief cause of his loneliness and consequent despair, this man can now use his self-knowledge to transform his circumstance. He might, for instance, use his newfound awareness to act more generously. As a result, he may find himself with more friends and consequently happier. Again, only humans possess this power of self-modification. As a matter of fact, it constitutes one of the most significant advantages of self-conscious awareness.

And it's not just our individual selves we have the capacity to transform but our entire species. With just one thought, one concept, one technology, any human, within his or her lifetime, can alter the course of the entire species. How much more versatile can a creature possibly get? Once again, knowledge is power with self-knowledge being perhaps the most potent knowledge of all, or, as the ancient Chinese philosopher Lao Tzu so eloquently expressed it, "Knowledge of others is intelligence; knowledge of self is wisdom. Mastery of others is strength; mastery of self is power."

If we are to accept this merger of Aristotelian and Socratic precepts, then we agree that the universal means of maximizing happiness/minimizing suffering lies in increasing our store of self-knowledge, that is, in learning as much as we possibly can about ourselves, both as individuals and as a species. Moreover, if a great deal of our behavior is guided by genetically inherited impulses, then in order to maximize our capacity for self-knowledge, and with it happiness, we must first seek to maximize our understanding of those inherent impulses that determine so much of what we do and think. Being that some biological impulses, particularly in their extremes, can lead us to potentially destructive behaviors, by learning to understand those impulses, we will be better equipped to master and contain them. Granted, no biological impulse can be completely eradicated. Nevertheless, by understanding the underlying nature of our biological impulses, we can try to channel some of their potentially hazardous energies into more productive outlets. As the behavioral geneticist Richard Dawkins expressed this same notion in his book *The Selfish Gene:*

> If you wish to build a society in which individuals cooperate generously and unselfishly towards a common good, you can expect little help from biological nature. Let us try to teach generosity and altruism, because we are born selfish. Let us understand what our own selfish genes are up to, because we may then at least have a chance to upset their designs, something that no other species has ever aspired to do.[121]

So what if it should turn out that human spirituality and religiosity are nothing more than the consequences of an inherited biological impulse? If indeed this is the case, shouldn't we at least inquire into the underlying nature of such an essential part of us?

As stated previously, no trait is perfect. Though each physical characteristic we possess provides us with some adaptive utility, each comes with its own drawbacks. Consequently, if spirituality and religiosity constitute inherent physical characteristics of our species, what might be some of their drawbacks? What negative impact might a spiritual or religious function have on our species? Only once we determine this will we be able to maximize this impulse's positive aspects while minimizing its negative. Once we begin to view spiritual and religious consciousness as evolutionary adaptations, only then will we be able to objectively determine the negative impact they might have on us and, from there, begin work on turning them into strengths.

Generally speaking, humankind's spiritual propensities are pretty harmless, just a means by which humans can temporarily abate some of the psychoemotional strain that comes as an inherent part of the human condition. It's really only when our spiritual sensibilities become bound up by some restrictive and dogmatic religious creed that problems arise. Consequently, I will focus my critique on the potential drawbacks of the religious impulse.

For all the advantages of possessing a religious instinct, for all the social cohesion it brings, the sense of community it fosters, and the alleged purpose and meaning it provides, religion has proven itself, time and again, to be a potentially hazardous impulse in us. As the philosopher Alfred North Whitehead expressed it:

> History, down to the present day, is a melancholy record of the horrors which can attend religion: human sacrifice, and in particular the slaughter of children, cannibalism, sensual orgies, abject superstition, hatred as between races, the maintenance of degrading customs, hysteria, bigotry, can all be laid at its charge. Religion is the last refuge of human savagery.[122]

Granted, none of the world's mainstream religions presently practice child sacrifice or cannibalism. Nevertheless, even with the proscription of such barbaric rites, religion continues to act as a divisive force, promoting discrimination and intolerance, inciting enmity, aggression, and war.

But why is it that the world's various religions, whose tenets are so often based on just and loving principles, so frequently find themselves so venomously pitted against one another, inciting such acts of hostility, aggression, and, at its worst, even genocide? Though every culture possesses the same inherent religious impulse, because each one emerges from its own unique historical and environmental circumstance, this same impulse is manifested differently in each culture. It is for this reason that so many different religions have emerged. Because each religion has faith that its beliefs–and only its beliefs–are representative of "the truth," the tenets and beliefs of each religion inherently contradict every other. For instance, if my God is true, how can yours be also? And if the laws and principles by which you abide are God's laws and principles, then of what consequence are mine? As a result of this unfortunate psychodynamic, each religious maintains an inherent antagonism for every other.

Moreover, our religious functions instill us with an inherent belief that we are immortal. Because each religion possesses its own unique interpretation of what immortality represents, each religion views every other as a threat to its notion of an immortal soul (i.e., "If my notion of heaven is true, how can yours be also?"). Consequently, each belief system perceives every other as a threat to its sense of immortality, and any threat to one's immortal soul is not something that any individual or society is likely to take lightly. As a result, our species tends to engage in what could be termed religious tribalism, a predisposition to justify territorial conquest in the name of one's Gods, a tendency that has marked our species' violent and bloody history.

Perhaps if we could learn to view religiosity as nothing more than a genetically inherited impulse, we'd be better able to contain its more destructive influences. If we could come to understand the underlying nature of this instinct, perhaps we could learn to temper the inevitable antagonism that each religion inherently feels for every other. If we were to recognize that our religiously generated fears and antipathies were merely the effects of an inherited impulse–as opposed to anything founded in reason–we might be able to curb this same impulse that has launched our species into a history of repeated religious war. How many more times must we justify acts of cruelty, murder, and genocide in the name of God and religion before we learn to tame this destructive impulse in us? Even in our present day, we need just look to the Middle East, India/Pakistan, Northern Ireland, Timor, and Serbia/Croatia–not to mention, as of 9/11, nearly the entire world–to witness the destructive grip the religious instinct has on our species.

Only once the human animal comes to terms with the fact that it has been born into a mental matrix–a neurological web of deceit–will we have a chance of offsetting this potentially destructive impulse in us. Knowledge is power, and it is high time that the science of spirituality and religiosity be made available to the world so that our species might see that there is another way. It is time that the study of spirituality and religiosity be taken out of the hands of philosophers, metaphysicians, and theologians and "biologized."

Not to suggest we should seek to eradicate religiosity altogether, but rather that we try to put it into scientific perspective. In itself, there is nothing wrong with the religious impulse in that it bonds us with our communities and, through faith, helps to reduce stress levels and bolster general health. It is rather the excesses of the religious impulse that represents the greatest threat. As a matter of fact, the excesses of nearly any impulse–be it for food, love, sex, or materials–can be potentially dangerous, if not lethal. In the case of the religious impulse, in its extreme, it fosters radical ideologies that promote dangerously discriminatory, fanatical, and martyristic behaviors.

During the time of our species' emergence, when humans lived in small nomadic tribes, perhaps it was necessary that we possess a religious impulse. At that time, religious consciousness provided us not just with a means to cope with anxiety and death, but also with a way to order and organize ourselves socially. Nevertheless, times have changed since then. Since our emergence, not only have humans successfully populated the planet but, in the process, have evolved from a species made up of small, closely-knit, isolated, nomadic communities into denizens of diverse civilizations.

Within a relatively short period of time, humans have transformed their environments into something very different from the ones into which they originally evolved. At the time of our emergence, we were little more than what Desmond Morris referred to as "naked apes," monkey people who lived in caves and could start fires and chip rocks. And look at us now, a mere hundred thousand years later (which is very little in terms of evolutionary time) living in concrete megalopolises and using advanced methods of energy, transportation, and communication. In essence, the physical conditions into which our species was initially selected have been drastically altered since the time of our inception. As a result, certain aspects of our inherent "hard-wiring" no longer suit our new conditions, rendering us an environmentally maladjusted species.

Perhaps during the dawn of Man, when humans had barely populated the planet and still lived in isolated communities, religious tribalism didn't represent the same threat it does today but rather helped to preserve a group's identity and consequent survival. As time passed, however, and our species grew in numbers, varying cultures with their numerous religions and ideologies began to expand into one another's territories, making religious tribalism an ever-increasing threat to the fabric of our new social arrangements. As the author Hermann Hesse expressed the same sentiment in somewhat harsher terms, "human life is reduced to real suffering–to hell–only when two ages, two cultures and religions overlap." Consequently, as we find ourselves living in what is an increasingly global community, maintaining a diversity of belief systems may no longer represent a viable option for our species. Instead, we may have to learn to adopt one unified set of religious and spiritual principles through which to achieve global harmony. Perhaps if we could learn to embrace a single humanistic ideology based on such principles as equality, tolerance, compassion, and forgiveness, we might be able to optimize our potential for happiness, while minimizing our potential for fostering pain and suffering in the world.

Humans are destined to remain a religious and spiritual animal. It is a fait accompli as we are "wired" this way. We therefore need to try to come up with practical solutions to deal with the problem of religious tribalism. One suggestion I would offer is that the leaders of the world's various religions should hold a consortium with the goal of writing up some sort of spiritual constitution, a book of universally accepted spiritual principles and guidelines by which each religion would agree to abide. For instance, should the world's religious leaders agree to endorse the basic ethic of "thou shalt not kill" (and under any circumstance), that alone would advance our species by leaps and bounds. Subsequently, should someone defy this religious constitution, he or she would be universally condemned as a terrorist and thereby deprived of any platform for their twisted cause. There is a U.N. in which the nations of the world seek mutual peace, cooperation, and stability; the world's religions are going to need to do the same. We should not underestimate or take for granted the galvanizing force religion has on people. Consequently, we need to hold religious institutions as accountable to upholding international law as we do of our world's nations.

Until we stop teaching our young to only honor and respect those with whom we share the same religious ideology, we are only encouraging the types of discriminatory values and behaviors that can only lead to our eventual mutual destruction. What else can come from generation after generation being brainwashed to believe that the lives of those outside their religious fold are less sacred than their own? The boundaries of respect for others must be extended beyond the narrow margins of any one religious paradigm and applied to the whole of humanity. Similar to the manner in which the Europeans have abandoned their national currencies and replaced them with one unified Euro, I'm suggesting that nations replace their religious ideologies with one, agreed upon, spiritual paradigm, one world religion based on a brotherhood of man. United, our species may have a chance of standing; divided, however, we are sure to eventually fall. As stated by Einstein in an impassioned plea to the nations of the world after our last world war, "Only a few short years remain in which to discover some spiritual basis for world brotherhood, or civilization as we now know it will certainly destroy itself."

This notion of containing our self-destructive impulses seems particularly relevant today in a world in which there's increasing availability of weapons of mass destruction. In such potentially precarious times, can we really afford to leave ourselves unchecked at the mercy of our most primal instincts? Just as it is necessary that we contain the excesses of all of our instincts, shouldn't we seek to do the same for our religious ones as well? Rather than to simply learn new ways to negotiate war, wouldn't we be better off if we sought to understand and thereby contain those impulses that continue to drive us to engage in one? There is no time left to negotiate. We've played our last chip in the war room. Any next world war that might emulate those of our past would mark the end of life as we know it. Again to quote Einstein, in all his eloquence, "I do not know what weapons will be used to fight World War III, but World War IV will surely be fought with sticks and stones."

Because our species is temporarily king of the hill, we presume that we're invincible. It's as if we've placed unconditional trust in the forces of nature to preserve us, as if, because of the great strength we presently possess, we are immune to the forces of extinction. Perhaps we feel this way because we continue to believe the myth that we are God's "chosen creatures." To recognize what a puerile fantasy such thinking represents, we need just look at terrestrial life's three-and-a-half-billion-year history to see that it is little more than a chronicle of mass extinctions. As a matter of fact, for every species that exists today, there are countless more that are now extinct.

Just because we happen to live in a time of relative peace and calm (if we can even say such a thing any longer), we shouldn't presume things will remain this way forever. The history of our species is an epic of war, one that is often contingent on the world's economic conditions which happen to be cyclical in nature, fluctuating between periods of growth and recession. In a period of growth, we become complacent. In recession, we go to war. Put a hundred loaves of bread before a hundred hungry people belonging to two different religions, and you will have peace. Put ten loaves of bread before a hundred hungry people of two different religions, and you will have genocide. And with all our newly advanced medical technologies, which decrease infant mortality rates and extend life expectancy, the continued rise in our world's population only exacerbates the possibility of a world recession.

In addition, because our religious functions compel us to believe in an afterlife, we allow ourselves to be profligate. Because we inherently perceive ourselves as immortal, we place less meaning and significance on perfecting ourselves within this lifetime as well as in preserving the conditions of this, our Earthly environment. Why, after all, worry about the Earth when we'll be spending the rest of eternity elsewhere? How else are we to explain the manner in which we recklessly continue to exploit and butcher this planet–as if we're the last living generation?

So why not use the same methodology (science) that has enabled us to master our environments to master ourselves? Isn't it time we begin placing the same emphasis we do on perfecting our toys–our spaceships, computers, and automobiles–into perfecting ourselves? How much longer will we be slaves to destructive religious creeds before we can transfer our faith over to the natural sciences? Why this need to cling onto the same antiquated paradigms by which we were raised? What if our great, great–and then some–grandparents were wrong? What if those who believed rain to be manna from heaven and lightning the wrath of God didn't know what they were talking about?

So which will it be? Are we to accept the underlying principles conceived in scientific method–in reason–or are we to obstinately hold on to those antiquated belief systems that sprang from our pre-scientific, ignorant past? In prior times, it was considered blasphemy to believe that the Earth revolved around the sun. Since such primitive times, science has sent men to the moon and back. In the past, it was considered sinful to perform an autopsy, to study human anatomy and physiology. Now, as a result of the physiological sciences, we've developed a plethora of medical technologies that have eased our pains and extended our life expectancies. And yet, in a society as modern as ours, in the world's most powerful democracy, we still find ourselves battling against the suppressive forces of religious ultraconservatism and fundamentalism. In as modern an age as ours, we still live in a nation in which the same evolutionary principles that brought us so many life-enriching technologies struggle to be taught in the classroom. And why? Because religious values, which so often seek to impede the march of scientific progress–of reason–continue to play a significant role in human nature and therefore in human politics.

We rely on our religions to tell us what is acceptable versus unacceptable, what we should and shouldn't do, what we can and can't say or think. Religion acts as a constricting force, constantly trying to obstruct the flow of any information it construes as a threat to its own obsolete ideology. In this way, religion confines us. It limits our field of vision. It tries to place us in a narrow box and bind us within it. Should we seek to step outside the confines of that box, to merely take a peek at the world of possibilities beyond, we are to be shunned and punished. Only why, when this life may be our last, should we want to limit ourselves in such a way?

Not to suggest that there should be no limits set on human behavior. As a social animal, with often runaway impulses, there's nothing wrong with a bit of healthy restraint. By no means am I encouraging the dissolution of all codes of conduct. It's just that do we necessarily want these codes to be based on antiquated mythologies? Through the careful application of scientific method we know more about the origins and nature of human behavior than ever before. Why then would we want to rely on systems that were based on the whims of men's imaginations, on untested and unproven hunches, to decide social doctrine? Should a person suffer from psychosis, should they seek the care of a licensed psychiatrist or an exorcist? Isn't it time we finally discard our dated paradigms and replace them with methods that can at least be validated? How much more evidence do we need before we will finally embrace the scientific process? And if we do, shouldn't we seek to resolve our social and ethical dilemmas through this same medium? As the sociologist Auguste Compte expressed it, "Only those willing to submit themselves to the rigorous constraints of scientific methodology and to the canons of scientific evidence should presume to have a say in the guidance of human affairs. Just as freedom of opinion makes no sense in astronomy or physics, it is similarly inappropriate in the social sciences."[123]

Suppose there is no spiritual reality. Suppose we are nothing more than strictly physical entities, a chance combination of molecules, devoid of any ghost in the machine. Granted, energy can neither be created nor destroyed. Granted, the same energy of which we are composed today will exist in some form until the end of time. Nevertheless, once our brain dies, once its cognitive processes stop functioning, so does our conscious experience. In whatever form our present store of energy will be redistributed into the vast universe after death, whether it be as soil, gas, or cosmic dust, it will bear no relation to who or what we are and experience today. Never again will we exist in the same exact molecular combination. Consequently, never again will we undergo the same conscious experience. As much as we would like to believe that we are somehow more than the sum of our physical parts, most likely we are not. It's therefore most likely that when the parts stop functioning, so does the whole. Whether we want to believe it or not, death is most likely the decisive end of personal identity. Being that this may therefore be our one and only shot at existence, shouldn't we seek to place our priorities and emphasis on actualizing ourselves here on Earth rather than putting all our hopes into some dubious hereafter?

Suppose we are composed of matter and nothing more. If true, we must learn to view ourselves as organic machines. Not until we accomplish this will we be able to effectively act as our own mechanics. If we truly possess a religious function, one that has instilled our species with an impulse leading us to acts of aggression, hostility, and war, shouldn't we seek to master it? If we truly are ticking biological time bombs, shouldn't we seek to diffuse ourselves?*

Besides, if there is no spiritual reality, just think of all the time and energy we've wasted in practicing our illusionary beliefs. Think of all of the useless rituals and ceremonies we've performed; the sacrifices we've made; the purses we've filled; the edifices we've built; the people we've oppressed, ostracized, beaten, and killed; the figments of our imaginations that we've bowed to and beseeched; and, meanwhile, all of it in vain. If there truly is no spiritual realm, we've been little more than "the absurd species" that has been wired to pay homage to thin air.

Imagine what a group of onlooking extraterrestrials would think after witnessing our behavior. "Look at the monkey people," they would say, "offering sacrifices to the void; killing, defiling, and warring with one another over literally nothing; banging their chests and wailing at the wind, all in the vain hope that it might incite some imaginary being to save them from their inevitable fates."

For the first time in our species' history, we possess a rational explanation of God. For the first time, we can justifiably dismiss our old religious and metaphysical paradigms as delusional impediments to progress and prosperity. Nietzsche may have hypothesized that God is dead, but science just confirmed it. Now that we can confidently dispel our old myths, let us dispose of those primitive ideologies that teach us to oppress women, freethinkers, and homosexuals,

---

*In light of the potentially hazardous nature of this impulse, one might ask: should we use future advances in the genetic sciences to eradicate the genes responsible for generating such divisive behaviors? Should we seek to strike religiosity from human consciousness forevermore? Considering the dangers of genetic tampering, I, for one, would not encourage such a drastic strategy. At the same time, I have heard others speak of the possibility of surgically removing one's "God" part of the brain as yet another option, a procedure that has been whimsically referred to as a Godectomy. As another solution, perhaps there will one day be pills that will help chemically suppress the excesses of this impulse in as much as we may one day view fanaticism as a type of "religious disorder" that requires medication. Regardless of these more intrusive solutions, if it's not already too late, we are probably more likely to resolve the problem of religious tribalism through the old-fashioned means of reason and diplomacy.

and that encourage us to discriminate against anyone taught a different set of fairy tales than we were. Let us unapologetically embrace a humanistic philosophy so that we can finally advance our social evolutions.

If it's true that there is no spiritual reality, no God, no soul, and no afterlife, then let's accept ourselves for what we are and make the most of it. Perhaps such a change in our self-perceptions might help us to shift our priorities from the hereafter to the here and now, to deter intolerance, antipathy, and war, thereby minimizing our pain and maximizing our chance of obtaining the greatest amount of happiness in life. This, more than anything, is what I would hope to gain from a scientific interpretation of human spirituality and God.

Let the secular revolution begin . . .

Epilogue

Quest's End

"We are not now that strength
which in old days
Moved Earth and heaven;
that which we are, we are;
One equal temper of heroic hearts,
Made weak by time and fate,
but strong in will
To strive, to seek, to find,
and not to yield."

–Alfred Lord Tennyson, *Ulysses*

"The key to achieving immortality is living a life worth remembering."

–St. Augustine

Here lies the end of my personal lifelong quest for knowledge of God. Though I'll always remain open to the possibility that a spiritual/transcendental realm might still exist, until that time, I trust in–that is, I have faith in–the solution I've provided for myself.

Granted, I would have preferred that my research yielded proof of a God, proof that there existed some transcendental realm through which "I," my conscious self, would have persisted forever. Sure, I would have preferred eternal existence over inevitable death. Or would I? Imagine the ramifications of immortality, of knowing that there will never be a moment's rest or respite from eternal existence.

Besides, amid eternity, what goals or motivations could one have? How relevant would anything be? Eventually, hours, years, eons would all blur together, rendering existence an endeavor in obscurity. It would be like being in a race with no finish–no winners, no losers, no anything...just existence for existence's sake. Under such conditions, what would prevent one from losing interest, from slowing down, from no longer pushing oneself to achieve? In such a light, what would achievement even mean? Perhaps it's better this way, better to burn quick and bright than forever dim. Perhaps without death, life would intrinsically lack luster and meaning. Perhaps so, perhaps not. Perhaps I'm simply trying to rationalize the subconscious fear of my inevitable demise.

So where to now? Knowing that I'm destined to grow old and infirm and eventually die, that I'm to lose everything I ever had or loved, including my own self, why, I sometimes ask, bother to continue living? Why, in a godless universe, should I continue to push this burdenous rock of Sisyphus just one more day? Why not just get it over with and kill myself right here and now? Though during some of the more distressing times in my life, I may sometimes toy with such ideas, I console myself with the realization that if there really is no spiritual realm, no soul, and no afterlife, then I'll have all eternity to not exist, to not have to endure the vagaries of capricious reality. With this in mind, why not make the most of this fleeting experience called life while it's still available to me? Even if I were to procure just one more moment of genuine happiness that would still be one more than nothing.

Perhaps the mere fact that we are cognizant of anything at all is reason enough to celebrate life. How many other combinations of matter can do what we can? What other molecular entity possesses the capacity to laugh; to love; to ponder its own existence; to appreciate works of music, art, literature; to aspire, to hope, to dream? Even if it should turn out that we are just spiritless atoms cavorting in the void, we are still matter's paramount form, the height of its complexity, its crème de la crème–nature's chosen macromolecules.

Besides, even if it should turn out that what we call happiness is nothing more than the manifestation of strictly physiological processes, do we experience it any less? Whether I'm mortal or immortal, a spiritual entity or a spiritless organic machine, are these not my experiences? Either way, am I any less *me*? Moreover, the mere fact that I can never know what each next moment will bring means that, as mechanical as life might be, mine remains a wondrous and beautiful mystery.

ADDENDUM

EXPERIMENTS THAT MIGHT HELP PROVE THE EXISTENCE OF A SPIRITUAL FUNCTION

"Scientific method today has reached about as far in its understanding of the human mind as it had in the understanding of electricity by the time of Galvani and Ampere. The Faradays and the Clerk Maxwells of psychology are still to come; new tools of investigation, we can be sure, are still to be discovered before we can penetrate much further, just as the invention of the telescope and calculus were necessary precursors of Newton's great generalizations in mechanics."

–JULIAN HUXLEY

"The truth will out!"

–SHAKESPEARE

1) Take ten highly spiritual and/or religious individuals from ten distinctly unique religious orientations (those from isolated cultures which practice a crude animism to technologically advanced western

cultures that engage in anything from organized religion to new age spiritualism) and submit them to a Functional Magnetic Resonance Imaging scanner (fMRI) while engaged in the act of prayer and/or spiritual contemplation. See if this produces similar effects in the neural activity of each of the participants.

1a) Conduct the same test on the same individuals, but instead of subjecting them to a fMRI, take blood from them to see if religious/spiritual activity might prompt any difference in their blood chemistry.

1b) Perform the same tests as above on a group of non-religious/atheistic individuals from different cultures and compare them to the results of the first group.

2) Take a group of one-year-olds. Perform a fMRI on them. Have them undergo similar scans once every year, until they reach the age of twenty. Once a site has been identified that represents the seat of spiritual cognition, look for changes to that site in each progressive scan that is taken on each subjects. In this way, we might be able to chart the development of the spiritual and religious functions in the human brain.

2a) In regard to the above fMRI readings, pay special attention to those individuals who undergo a religious conversion. Compare the scan results of those who have undergone conversions, not only to their old scans (before they converted), but also to those who haven't converted at all.

3) Once a site has been identified as the seat of spiritual and/or religious consciousness, study cases of individuals who have either had that part surgically removed or who have suffered some sort of damage to that area (e.g., a stroke or head trauma) and see to what degree, if any, this may have affected these individuals' spiritual sensibilities and/or religious attitudes and behaviors. Such tests should confirm whether or not it is possible for humans to suffer from spiritual or religious aphasias.

ENDNOTES

1. William Keeton, *Biological Science* (W. W. Norton and Company, Inc., 1980), 896.
2. Ibid A8.
3. Ibid 65.
4. Ibid 491.
5. Yoshiya Asano et al., "Rhodopsin-like proteins in planarian eye and auricle: detection and functional analysis," *Journal of Experimental Biology*.
6. R. A. Hinde, *Biological Bases of Human Social Behavior* (New York: Mcgraw-Hill, 1974), 38.
7. William Keeton, *Biological Science* (W. W. Norton and Company, Inc., 1980), 492.
8. Ralph Linton, *Science of Man in the World Crisis* (New York: Octagon Books, 1978), 123.
9. John Blacking, *How Musical Is Man?* (Faber & Faber, 1976), 7.
10. Anthony Storr, *Music and the Mind* (Ballantin, 1992), 1.
11. Ibid 29.
12. Ibid 35.
13. Ivar Lissner, *Man, God and Magic* (New York: Putnam, 1961), 12.
14. E. O. Wilson, *On Human Nature* (New York: Bantam Books, 1976), 176.
15. Dr. Herbert Benson, *Timeless Healing* (Scribner, 1996), 198.
16. Carl Jung, *Collected Works*, vol. 9 Part 1 4–5.
17. Frieda Fordham, *An Introduction to Jung's Psychology* (New York: Penguin Books, 1953), 70.
18. Mircea Eliade, *The Sacred and the Profane* (Harcourt Brace Jovanovich, 1959), 11.
19. E. Heobel and E. Frost, *Cultural and Social Anthropology*, 348.
20. Bronislaw Malinowski, "The Group and the Individual in Functional Analysis," *American Journal of Sociology* 44 (May 1939): 959.
21. *Encyclopædia Britannica*, 15th ed., 127.
22. Mircea Eliade, *The Sacred and the Profane* (Harcourt Brace Jovanovich, 1959), 87.
23. Anthony Steven, *On Jung* (Routledge, 1990), 143.
24. E. O. Wilson, *On Human Nature* (New York: Bantam Books, 1976), 1.
25. Robin Fox, *The Cultural Animal*, 273–96.

26. Raj Persaud, "God's in Your Cranial Lobes," *Financial Times* (May 8–9 1999).
27. Sigmund Freud, *Civilization and Its Discontents* (W. W. Norton and Co., Inc, 1962), 25.
28. Ernest Becker, *Denial of Death* (The Free Press, 1973), 17.
29. Ralph W. Hood Jr. et al., *The Psychology of Religion* (The Guilford Press, 1996), 153.
30. G. Zilboorg, "Fear of Death," *Psychanalytic Quarterly* (1943): 12:465–67.
31. *Encyclopædia Britannica*, 15th ed., 201.
32. Dr. Herbert Benson, *Timeless Healing* (Scribner, 1996), 198.
33. Sigmund Freud, *The Future of an Illusion* (New York: Norton, 1927), 22.
34. Sigmund Freud, *Civilization and its Discontents* (W. W. Norton and Co., Inc, 1962), 20.
35. Ralph W. Hood Jr. et al., *The Psychology of Religion* (The Guilford Press, 1996), 161.
36. M. Ostow and B. A. Scharfstein, *The Need to Believe* (International University Press, 1953), 23.
37. Karen Armstrong, *A History of God: The 4,000 Year Quest of Judaism, Christianity and Islam* (New York: Knopf, 1993), 208.
38. Sigmund Freud, *Civilization and Its Discontents* (W. W. Norton and Co., Inc., 1962), 11.
39. Ibid 21.
40. Dan Merkur, *Gnosis: An Esoteric Tradition of Mystical Visions* (Albany, NY: State University of New York Press, 1993), 8.
41. Ibid 9.
42. Albert Einstein, *Ideas and Opinions* (New York: Crown Publishers, 1954), 64.
43. S. Freud, *Civilization and Its Discontents* (W. W. Norton and Co., Inc., 1962), 12.
44. R. W. Hood Jr., *Mysticism*, 285–297.
45. R. K. Forman, *The Problem of Pure Consciousness*, 8.
46. R. M. Bucke, *Cosmic Consciousness: A Study of the Evolution of the Human Mind* (University Books, 1961), 67.
47. M. M. Poloma and B. F. Pendleton, *Review of Religious Research* (1989), 48.
48. Savage, Hoffman, Fadiman, and Savage, 1971.
49. J. Jaynes, *The Origin of Consciousness in the Breakdown of the Bicameral Mind*, 360.
50. R.D. Laing, from Ralph Metner's *The Ecstatic Experience*, 15.
51. Wilson, Elgin, Vaughan, and Wilber, "Paradigms in Collision" from *Beyond Ego: Transpersonal Dimensions in Psychology*, 47.
52. Ibid 47.
53. Daniel Goleman, "A Map for Inner Space" from *Beyond Ego*, 147.
54. C. D. Batson and W. L. Ventis, *The Religious Experience* (Oxford University Press, 1982), 98.
55. M. Pafford, *Inglorious Wordsworths*, 262.
56. R. Walsh, D. Elgin, F. Vaughan, and K. Wilber, "Paradigms in Collision" from *Beyond Ego*, 41.
57. R. Stark, "A Taxonomy of Religious Experience" (*Journal for the Scientific Study of Religion*, 5, 1965), 165–176.
58. W. James, *Varieties of the Religious Experience*, 315.

59. Woodruff (1993) Report: Electroencephelograph taken from Pastor Linton Pack, In T. Burton, "Serpent-Handling Believers," 142–144.
60. Stanislav Grof, *Realms of the Human Unconscious* (The Viking Press, 1975), 204.
61. J. Blofeld, *The Tantric Mysticism of Tibet*, 24.
62. Ralph W. Hood Jr. et al., *The Psychology of Religion* (The Guilford Press, 1996), 229.
63. Kaplan and Sadock, *The Comprehensive Textbook of Psychiatry*, 7th ed., 445.
64. Soren Kierkegaard, *Sickness Unto Death*, 146.
65. V. S. Ramachandran, *Phantoms in the Brain* (New York: William Morrow & Co., Inc., 1998), 250.
66. Ibid 252.
67. Ibid 225.
68. S. Freud, *Civilization and its Discontents* (W. W. Norton and Co., Inc., 1962), 14.
69. Norman O. Brown, *Life Against Death* (Vintage Books, 1959), 159.
70. S. Freud, *Civilization and its Discontents* (W. W. Norton and Co., Inc., 1962), 16.
71. B. K. Anand, G. S. Chhina, and B. Singh, "Electroencephalography and Clinical Neurophysiology" 13, 1961, 452–456.
72. Steven Rose, *The Conscious Brain* (Vintage Books, 1976), 335.
73. Batson and Ventis, *The Religious Experience* (Oxford University Press, 1982), 98.
74. Robert Jesses "Entheogens: A brief history of their Spiritual Use" (*Tricycle*: Volume 6, Number One: Fall 1996), 60.
75. Ibid 62.
76. Freud, *Civilization and its Discontents* (W. W. Norton and Co., Inc., 1962), 13.
77. M. Peyser and A. Underwood, "Shyness, Sadness, Curiosity, Joy. Is it Nature or Nurture?" *Newsweek*, Special spring/summer edition, 1998, 62.
78. M. Nash, "Fertile Minds," *Time*, Feb 3, 1997, 50.
79. Sharon Begley, "How to Build a Baby's Brain," *Newsweek*, Special spring/summer edition, 1998, 30.
80. M. Peyser and A. Underwood, "Shyness, Sadness, Curiosity, Joy. Is it Nature or Nurture?" *Newsweek*, Special spring/summer edition, 1998, 62.
81. *Journal of Personality*: 67:6, December, 1999, 957.
82. Ibid 962.
83. Ibid 952.
84. Ibid 952.
85. Ibid 964.
86. Dr. Raj Persaud, *Financial Times* (May 8/May 9, 1999), X Weekend FT.
87. Benson, *Timeless Healing* (Scribner, 1996), 157.
88. Ibid 157.
89. W. James, *Varieties of the Religious Experience*, 162.
90. Ralph W. Hood Jr. et al., *The Psychology of Religion* (The Guilford Press, 1996), 279.
91. Ibid 117.
92. S. De Sanctis, *Religious Conversion: A Bio-Psychological Study*, 67.
93. Ralph W. Hood Jr. et al., *The Psychology of Religion* (The Guilford Press, 1996), 289.
94. Ibid 280.

95. Pratt, J.B., *The Religious Consciousness: A Psychological Study* (New York: MacMillan, 1920), 113.
96. Ostow and Scharfstein, *The Need to Believe* (International University Press, 1953), 102.
97. Ralph W. Hood Jr. et al., *The Psychology of Religion* (The Guilford Press, 1996), 279.
98. Ibid 399.
99. *International Journal for the Psychology of Religion*: 2000, 10(3), 185.
100. *Journal of Neuropsychiatry*: Clinical Neuroscience; 1997, Summer; 9(3):498.
101. K. L. R. Jansen, M.D., *Using Ketamine to Induce the Near-Death Experience*, 64.
102. Ibid 73.
103. Diana Eck, *A New Religious America: How a "Christian Country" Has Become the World's Most Religiously Diverse Nation* (Harper San Francisco, 2001).
104. Zorach v. Clauson, 343 U.S. 306, 313 1952.
105. Phil Zuckerman, *Atheism: Contemporary Rates and Patterns, Cambridge Companion to Atheism*, ed. Michael Martin (Cambridge University Press, 2005).
106. Phil Zuckerman, *Atheism: Contemporary Rates and Patterns, Cambridge Companion to Atheism*, ed. Michael Martin (Cambridge University Press, 2005).
107. D. T. Suzuki, A. J. F. Griffiths, J. H. Miller, and R. C. Lewontin, in *An Introduction to Genetic Analysis* 4th ed. (W. H. Freeman, 1989).
108. L. B. Jorde et al., "Gene Mapping in Isolated Populations," *Journal of Human Heredity*; 2000; 50:57-65.
109. *Encyclopaedia Britannica*, 15th Edition, Volume 18, 924.
110. Richard Current et al., *American History*, 7th Edition (New York: Alfred A. Knopf, 1987), 224.
111. *Encyclopaedia Britannica*, 15th Edition, Volume 18, 928.
112. E. O. Wilson, *Sociobiology* (London: Belknap Press; Harvard University Press, 1980), 287.
113. Rob Stein, *Sociality, Morality and the Brain*, Monday, October 25, 1999; A13.
114. Ibid.
115. Reuters Health, "Researchers Identify Brain's Moral Center," Wednesday (5/3/00).
116. Reuters Health, "Researchers Identify Brain's Moral Center," Wednesday (5/3/00).
117. Nicholas Regush, *The Breaking Point* (Toronto: Key Porter Books Limited, 1997), 101.
118. Ibid 102.
119. Ralph W. Hood Jr. et al., *The Psychology of Religion* (The Guilford Press, 1996), 19.
120. S. Aurobindo, *The Future Evolution of Man* (All India Press, 1963), 27.
121. Richard Dawkins, *The Selfish Gene*, 82.
122. A. N. Whitehead, *Religion in the Making* (Macmillan, 1926), 37.
123. Coser, A. L., *Masters of Sociological Thought* (Harcourt Brace Jovanovich, 1997), 5.

BIBLIOGRAPHY

American Psychiatric Association. *Diagnostic and Statistical Manual of Mental Disorders.* (4th ed.) Washington, D.C.: American Psychiatric Association, 1994.

Armstrong, Karen. *A History of God: The 4,000 year quest of Judaism, Christianity and Islam.* New York: Knopf, 1993.

Aurobindo, S. *The Future Evolution of Man.* All India Press, 1963.

Barrett, David, George Kurian, and Todd Johnson in *World Christian Encyclopedia* (2001). Oxford University Press.

Bateson, G. *Mind and Nature.* E.P. Dutton, 1979.

Batson, C. D. and W. L. Ventis. *The Religious Experience.* Oxford University Press, 1982.

Becker, E. *Denial of Death.* The Free Press, 1973.

Benedict, R. *Patterns of Culture.* Houston Mifflin, 1989.

Benson, H. *The Relaxation Response.* New York: Morrow,1975.

Benson, H. *Timeless Healing.* Scribner, 1996.

Blacking, J. *How Musical is Man?* Faber & Faber, 1976.

Bootzin, R., J. R. Acocella, and L. B. Alloy. *Abnormal Psychology.* McGraw-Hill, 1993.

Brooks, J. S. and T. Scarano. "Transcendental meditation in the treatment of post-Vietnam adjustment." *Journal of Counseling and Development* (1985): 65:212-215.

Brown, N. O. *Life Against Death.* Vintage Books, 1959.

Bucke, R. M. *Cosmic Consciousness: A Study of the Evolution of the Human Mind.* University Books, 1961.

Campbell, J. *Transformations of Myth Through Time.* Harper and Row, 1990.

Cavendish, R. *Mythology.* Rizzoli International Publications Inc., 1980.

Chance, M. R. A. "Social Behavior and Primate Evolution." In M.F. Ashley Montagu, ed. *Culture and the Evolution of Man.* Oxford University Press, 1962.

Choron, J. *Death and Western Thought.* Collier Books, 1963.

Clark, W. H. *Chemical Ecstasy: Psychedelic Drugs and Religion.* Sheed and Ward, 1969.

Coren, S., C. Porac, and L. Ward. *Sensation and Perception.* New York: Academic Press, 1978.

Cohen, D. *The Circle of Life: Rituals From the Human Family Album.* Harper San Francisco, 1991.

Cooper, M., and M. Aygen. "Effect of meditation on blood cholesterol and blood pressure." *Journal of the Israel Medical Association* (1978): 95:1-2.

Coser, A. L. *Masters of Sociological Thought: Ideas in Historical and Social Context.* Harcourt Brace Jovanovich, 1997.

Davidson, J., and R. Davidson. *The Psychology of Consciousness.* Plenum, 1979.

Davie, Grace. *Religion in Modern Europe.* Oxford University Press, 2000.

Dobzhansky, T. "Anthropology and the Natural Sciences–the Problem of Human Evolution." *Current Anthropology* (1963): 4:138, 146-148.

D'Onofrio, B., L. Eaves, L. Murrelle, H. Maes, and B. Spilka. (67;6) *Journal of Personality: Understanding Biological and Social Influences on Religious Affiliation, Attitudes, and Behaviors: A Behavioral Genetic Perspective.* Blackwell, 1999.

Durant, W. *The Story of Philosophy.* Washington Square Press, 1961.

Eccles, J.C. *The Neurophysiological Basis of Mind.* Clarendon Press, 1953.

Eck, Diana. *A New Religious America: How a "Christian Country" Has Become the World's Most Religiously Diverse Nation.* Harper San Francisco, 2001.

Eliade, M. *The Encyclopedia of Religion.* MacMillan, 1987.

Eliade, M. *The Sacred and The Profane: The Nature of Religion.* Harcourt Brace Jovanovich, 1959.

Einstein, Albert. *Ideas and Opinions.* New York: Crown Publishers, 1954.

Etcoff, Nancy. *Survival of the Prettiest: The Science of Beauty.* New York: Random House, 1999.

Fordham, F. *An Introduction to Jung's Psychology.* New York: Penguin Books, 1953.

Forman, R. K. *The Problem of Pure Consciousness.* Oxford University Press, 1980.

Freud, S. *Civilization and Its Discontents.* W. W. Norton and Co., Inc., 1962.

Freud, S. *The Future of an Illusion.* J. Strachey, Trans. New York: Norton, 1927.

Freud, S. *Inhibitions, Symptoms and Anxiety.* New York: Norton, 1926.

Froese, Paul. "After Atheism: An Analysis of Religion Monopolies in the Post-Communist World." *Sociology of Religion* (2004): 65:57–75.

Fromm, E. *The Sane Society.* Fawcett Publications, 1955.

Foucalt, M. *Madness and Civilization.* New York: Random House, 1965.

Fox, Robin. *The Cultural Animal.* Smithsonian Institution Press, 1971.

Furst, P. *Flesh of the Gods: The Ritual Use of Hallucinogens.* Praeger Publishers, 1972.

Gallup, George, and Michael Lindsay. *Surveying the Religious Landscape.* Harrisburg, PA: Morehouse Publishing, 1999.

Gallup, G. G., Jr. "Chimpanzees: Self-recognition." *Science* (1970): 167: 86–7.

Gallup, G. G., Jr. "Mirror-image stimulation." *Psychological Bulletin.* (1972): 70: 782–93.

Geschwind, N. "Language and the brain." *Scientific American* (1972): 226: 76–83

Goleman, D. *The Varieties of the Meditative Experience.*, New York: Irvington Publishers, 1977.

Goleman, D., and R. Davison. *Consciousness: Brain, States of Awareness and Mysticism.* Harper and Row, 1979.

Greeley, Andrew. *Religion in Europe at the End of the Second Millennium.* New Brunswick, NJ: Transaction Publishers, 2003.

Greyson, B. "The Psychodynamics of a Near-Death Experience." *Journal of Nervous and Mental Disease.* (1983): 376–80

Grof, S. *Realms of the Human Unconscious.* The Viking Press, 1975.

Hamilton, W. D. "The Genetical Theory of Social Behavior." I,II. *Journal of Theoretical Biology* (1964): 7:1–52.

Hamilton, W. D. "Selfish and Spiteful Behavior in an Evolutionary Model" (1970) 228:1218–20.

Heelas, P. *Social Anthropology and the Psychology of Religion.* Elsmsford, NY: Pergamon Press, 1985.

Heobel, E., and E. Frost. *Cultural and Social Anthropology.* McGraw-Hill Book Company, 1976.

Hill, P. C. *Affective Theory and Religious Experience.* Birmingham AL: Religious Education Press, 1995.

Hinde, R. A. *Animal Behavior.* New York: McGraw-Hill Press, 1970.

Hinde, R. A. *Biological Bases of Human Social Behavior.* New York: McGraw-Hill Press, 1974.

Hood, R. W. Jr., B. Spilka, B. Hunsberger, and R. Gorsuch. *The Psychology of Religion.* The Guilford Press, 1996.

Hutchison, R. William. *Religious Pluralism in America: The Contentious History of a Founding Ideal.* New Haven: Yale University Press, 2003.

Ingelhart, Ronald, ed. *Human Values and Social Change.* Boston, MA: Brill, 2003.

Jansen, K. L. R. "Transcendental Explanations and the Near-Death Experience." Lancet. (1991): 337, 207–43.

Jansen, K. L. R. "Non-Medical use of Ketamine." *British Medical Journal.* (1993): 298, 4708–4709.

Jansen, K. L. R. *Neuroscience, Ketamine and the Near-Death Experience; The Role of Glutamate and the NMDA Receptor in the Near-Death Experience.* New York: Routledge,1995.

Jansen, K. L. R. "The Ketamine Model of the Near-Death Experience: A Central Role for the NMDA Receptor." *Journal of Near-Death Studies* (1996) ed. B. Greyson.

Janssen, J., J. de Hart, and M. Gerhadts "Images of God in Adolescence." *International Journal of the Psychology of Religion.* (1994): 4, 105–121.

Johnstone, Patrick. *Operation World.* Grand Rapids, MI: Zondervan Publishing House, 1993.

Jorde, L. B., W. S. Watkins, J. Kere, D. Nyman, and A. W. Eriksson. "Gene Mapping in Isolated Populations." *Journal of Human Heredity* (2000): 50:57–65.

Joselit Weismann, Jenna. *Immigration and American Religion.* Oxford University Press, 2001.

Jung, C. G. *The Undiscovered Self.* Mentor, 1958.

Jung, C. G. *Man and His Symbols, Psychology of Religion, and Symbolic Life.* Translated by R. F. C. Hull. Bollingen Series XX. Princeton Univ. Press. 1967–1978.

Jung, C. G. *The Portable Jung.* Edited by Joseph Campbell. Translated by R.F.C. Hull. New York: Penguin Books, 1976.

James, W. *Varieties of the Religious Experience.* Collier Books, 1902.

Jennings, J. George. "An Ethnological Study of Glossolalia." *Journal of the American Scientific Affiliation* (March 1968).

Jesses, R. "Entheogens: A brief history of their Spiritual Use." *Tricycle* (1996): 6:1.

Jones, W. L. *A Psychological Study of Conversion.* London: Epworth, 1937.

Kabat-Zinn, J., L. Lipworth, et al. "Four-year follow-up of a meditation-based program for the self-regulation of chronic pain." *Clin. J. Pain* (1986): 2:150–173.

Katz, S. T. *Mysticism and Language.* Oxford University Press, 1992.

Kedem, Peri. "Dimensions of Jewish Religiosity." *Israeli Judaism,* ed. Shlomo Deshen, Charles Liebman, and Mishe Shokeid. London: Transaction Publishers, 1995.

Koenig, H. G. *Aging and God: Spiritual Pathways to Mental Health in Midlife and Later Years.* New York: Haworth Press, 1994.

Kierkegaard, S. *Sickness Unto Death.* Princeton University Press, 1974.

Keeton, William. *Biological Science.* W. W. Norton and Company, Inc., 1980.

Kohlberg, L. *Essays on Moral Development: Vol. 2.* "The Psychology of Moral Development: The Nature and Validity of Moral Stages." San Francisco: Harper and Row, 1984.

Kolb, B., and I. Q. Whishaw. *Fundamentals of Human Neuropsychology.* (3rd ed.). New York: W.H. Freeman, 1990.

Lambert, Frank. *The Founding Fathers and the Place of Religion in America.* Princeton University Press, 2003.

Leary, T. F. *Flashbacks, an Autobiography.* J. P. Tarcher. L.A., 1983.

Lee Hotz, Robert. "Seeking the Biological Origins of Spirituality." *Los Angeles Times* (April 26, 1998).

Le Doux, Joseph. "Emotion, Memory and the Brain." *Scientific American* (1994): 270: 32–9.

Le Doux, Joseph. *The Emotional Brain.* New York: Simon & Schuster, 1996.

Leiman, A., and M. Rosenzweig. *Physiological Psychology.* D.C. Heath and Company, 1982.

MacKenzie, N. *Dreams and Dreaming.* London: Aldus Books, 1965.

Mahesh Yogi, M. *Transcendental Meditation.* New American Library, New York, 1963.

Malinowski, B. "The Group and the Individual in Functional Analysis," *American Journal of Sociology* (1939): 959,44.

Martin, W. T. "Religiosity and United States Suicide Rates, 1972–1978." *Journal of Clinical Psychology* (1984): 40, 1166-1169

May, R. *The Meaning of Anxiety.* Ronald Press, 1950.

Mayr, E. *Animal Species and Evolution.* Harvard University Press, 1963.

Melton, J. G. *Encyclopedia of American Religion.* 6th ed. Dale Publishing, 1999.

Merkur, Dan. *Gnosis: An Esoteric Tradition of Mystical Visions and Unions.* Albany, NY: State University of New York Press, 1993.

Morris, D. *The Naked Ape.* Jonathan Cape, 1969.

Morris, D. *The Human Zoo.* Jonathan Cape, 1970.

Noelle, C. David. "Searching for God in the Machine," *Free Inquiry,* Summer 1998.

Ostow, M., and B. A. Scharfstein. *The Need to Believe.* International University Press, 1953.

Noll, Mark. *Religion and American Politics: From the Colonial Period to the 1980s.* Oxford University Press, 1989.

Parrinder, G. *World Religions: From Ancient History to the Present.* Facts on File Publications, 1971.

Persaud, Raj, "God's in your Cranial Lobes," *Financial Times,* May 8/May 9 1999.

Persinger, Michael. *Neuropsychological Bases of Belief.* New York: Praeger, 1987.

Pfeiffer, F. *The Emergence of Man.* McGraw-Hill, 1977.

Pratt, J. B. *The Religious Consciousness: A Psychological Study.* New York: MacMillan, 1920.

Prince, R. H. *Religious Experience and Psychopathology.* Oxford University Press, 1992.

Pruyser, P. W. *A Dynamic Psychology of Religion.* New York: Harper, 1968.

Pugh, G. *The Biological Origin of Human Values.* Basic Books, 1977.

Ramachandran, V. S. *Phantoms in the Brain: Probing the Mysteries of the Human Mind.* New York: William Morrow & Co., Inc., 1998.

Rank, O. *Psychology and the Soul.* Perpetua Books, 1961.

Regush, Nicholas. *The Breaking Point: understanding your potential for violence.* Toronto, Canada. Key Porter Books Limited, 1997.

Reynolds, F. E., and E. H. Waugh. *Religious Encounters with Death: Insights from the History and Anthropology of Religion.* Pennsylvania State University Press, 1977.

Richmond, P. G. *An Introduction to Piaget.* Routledge and Kegan Paul, 1970.

Rose, Steven. *The Conscious Brain.* Vintage Books, 1976.

Rosenbleuth, A. *Mind and Brain.* The M.I.T. Press, 1970.

Sacks, O. *Awakenings.* Duckworth, 1973.

Sadock, Benjamin, and Virginia Sadock. *The Comprehensive Textbook of Psychiatry*, 7th ed. Lippincott, Williams and Wilkins, 2000.

Shand, Jack. "The Decline of Traditional Christian Beliefs in Germany." *Sociology of Religion* (1998): 59(2):179–184.

Sputz, R. "I Never Met a Reality I Didn't Like: A Report on 'Vitamin K'." *High Times* (October 1989): 64–82.

Stack, S., and I. Wasserman. "The Effect of Religion on Suicide." *Journal for the Scientific Study of Religion* (1992): 31, 457–466.

Stark, R. "A Taxonomy of Religious Experience." *Journal for the Scientific Study of Religion* (1965): 5.

Stevens, Anthony. *On Jung.* Routledge, 1990.

Storr, A. *Music and the Mind.* Ballantin, 1992.

Suzuki, D. T., A. J. F. Griffiths, J. H. Miller, and R. C. Lewontin in *An Introduction to Genetic Analysis.* 4th ed. W. H. Freeman, 1989.

Tillich, Paul. *The Courage to Be.* Yale University Press, 1952.

Turnbull, C. *The Human Cycle.* Simon and Schuster, 1983.

Turner, V. *The Ritual Process.* Aldine Publishing Company, 1969.

Van Gennep, A. *The Rites of Passage.* The University Of Chicago Press, 1960.

Walsh, R., and F. Vaughan. *Beyond Ego: Transpersonal Dimensions in Psychology.* J.P. Tarcher, Inc., 1980.

Whitehead, A. N. *Religion in the Making.* Macmillan, 1926.

Williams, Peter. *America's Religions: From Their Origins to the Twenty-First Century.* 2nd ed. University of Illinois Press, 2001.

Wilson, E. O. *On Human Nature.* New York: Bantam Books, 1976.

Wilson, E. O. *Sociobiology.* London: Belknap Press; Harvard University Press, 1980.

Young, J. Z. *A Model of the Brain.* Clarendon Press, 1964.

Zilboorg, G. "Fear of Death." *Psychanalytic Quarterly*, (1943): 12:465–467.

Zuckerman, Phil. "Atheism: Contemporary Rates and Patterns." *Cambridge Companion to Atheism.* Ed. Michael Martin. Cambridge University Press, 2005.

INDEX

H

I

About the Author

From early childhood–when he first realized he was one day going to die–Matthew Alper set himself upon a life journey to ascertain whether or not there exists a spiritual reality, a God. Was he merely a flesh-and-bone mortal or something more, something that perhaps transcended the restrictions of his fragile and very mortal physical existence? After receiving a BA in philosophy, Matthew continued his unconventional journey, working as everything from a photographer's assistant in New York City to a fifth-grade and high school history teacher in the projects of Brooklyn, a truck smuggler in Central Africa, and a screenwriter in Germany, and then he returned to New York City where he wrote what he considers his life's work, *The "God" Part of the Brain.* Since its initial publication in 1996, Matthew has lectured across the United States, appeared on NBC, done numerous radio shows, had his book used by numerous colleges to teach a variety of courses, and been praised by Pulitzer Prize–winners and other preeminent scholars and scientists. He is a contributor to the anthology *Neurotheology,* the emergent new science of which he is considered one of the founders. He presently lives in Park Slope, Brooklyn, with his cat, Sucio.